Battery Power Management
for Portable Devices

For a complete listing of titles in the
Artech House Power Engineering Series,
turn to the back of this book.

Battery Power Management for Portable Devices

Yevgen Barsukov

Jinrong Qian

ARTECH HOUSE

BOSTON | LONDON

artechhouse.com

Library of Congress Cataloging-in-Publication Data
A catalog record for this book is available from the U.S. Library of Congress.

British Library Cataloguing in Publication Data
A catalogue record for this book is available from the British Library.

Cover design by Vicki Kane

ISBN 13: 978-1-60807-491-4

© 2013 ARTECH HOUSE
685 Canton Street
Norwood, MA 02062

10 9 8 7 6 5 4 3 2 1

Contents

v

Preface

When holding an iPhone or some other marvel of modern electronics in your hand, it is hard not to wonder how engineers were able to put so much functionality into such a small space. While the complete answer to this mystery is that it is a pinnacle of thousands of years of human ingenuity, the simplified answer could be that it is a product of integrated circuits (ICs) that shrink in size twice every year and an amazing new power source—the lithium ion battery and its battery power management technologies.

For an engineer or a curious do-it-yourself enthusiast, it is not enough to know what kind of "magic" makes something tick. You also want to use the same "magic" ingredients to make your own incredible new devices. The purpose of this book is to make clear the batteries and battery management electronics that are used to operate in the most efficient and safest way.

We begin by introducing you to the full palette of modern rechargeable batteries. The book begins with rechargeable battery fundamentals, explaining how any battery operates and describing the common factors to consider when choosing a battery. Then we go into details on each of the popular chemistries. Along with describing each chemistry's peculiarities, answers to many practical common questions important information about rechargeable batteries is provided, including each type of battery's optimal charging procedure, energy and power capabilities, main applications, and cycle life, as well as battery chemistry safety.

Why address the question of safety? Because a typical battery pack has one-third of the energy of a hand grenade. It is this serious amount of power that enables battery-powered devices to achieve amazing functionality, but such power needs to be used with caution. After all, we are giving these devices to children, transporting them in airplanes, implanting them inside our bodies, and sometimes leaving them in hot cars (a bad idea!). When we design a device, it is our responsibility to make sure that it will work in all environments and will not turn into a fireball spitting out flaming projectiles. The topic of safety both in theory and implementation has the highest emphasis throughout this book, as both authors have spent a large part of their careers at Texas Instruments (TI), which produces most of the world's battery safety ICs.

Every battery has a finite amount of energy. Is there anything we can do to make a device run longer on that amount? It turns out, there is. One nonobvious point is that most devices we deal with are data-processing devices, which means that, at any point in time, a device may be holding precious unique information that would be lost forever if it were to suddenly shut down. You would not want to lose a PowerPoint presentation that you had been working on for the past hour or a hilarious e-mail that you were about to send. For this reason, we need to ensure that data-processing systems will save all data and go into sleep mode before a battery runs out of energy; that is, they must perform a *soft shutdown*. Here is the dilemma: If the system shuts down too early, the run time of the battery is reduced and, hence, you are not fully using the battery's energy! If the system shuts down too late, you lose the data. For example, if you know with 90% accuracy when battery voltage will drop below system shutdown voltage, you will have to shut the system down 10% earlier—hence, 10% of the run time is lost! This means that the error of the system remaining run-time estimation equals the loss of our system run-time. For this reason, the best handheld devices have a sophisticated capacity gauging system. More than half of all popular handheld systems, such as smartphones or laptops, are using TI Impedance Track™ technology, which allows for 99% accuracy under typical operating conditions. In this book, we explain the principles of battery gauging and give practical instructions on how to use it in your own devices.

One of the first issues encountered by anyone designing a system with a battery is that a battery needs to be charged. Each battery chemistry has its own charging algorithm, and some have many options. Do you want to charge it in the fastest possible way and also maximize capacity? Or do you want the battery to last as long as possible? Do you need to know how to charge a battery that has many cells in series, such that each cell has a different state of charge? Do you want to go with the cheapest or the most energy-efficient solution? Do you want to design the battery charging system so that it charges the battery while also powering the system? All of these questions will be discussed in a systematic way in the battery charger and cell balancing sections.

Finally, we combine all of the various pieces of knowledge about battery management and provide examples of complete battery-operated systems that can be cut-and-pasted into your own designs. After reading this book, we suspect you will emerge with a completely new appreciation of the complexity of both batteries and their power management systems. You will also be armed with sufficient knowledge to design such systems and to surprise us with new additions to our essential device's vocabulary that will transform everybody's lives.

Acknowledgments

Some of the material in this book has evolved from battery management industry seminars and deep diving tutorials conducted during the past eight years at Texas Instruments. We are very grateful for the excellent support from Texas Instruments for allowing us to share the advanced battery management technologies with the battery management industry and its various communities. We highly appreciate the following colleagues for their suggestions and support: Bill Jackson, Sihua Wen, Doug Williams, Ming Yu, Dave Freeman, Dave Heacock, Gaurang Shah, and Steve Lambouses.

Jinrong Qian would also like to thank his wife, Chun Lin, and his lovely son and daughter, Robert Qian and Jennifer Qian, for understanding and sacrificing their weekends and holidays.

Yevgen Barsukov would like to thank his wife and muse, Sudeshna Day, for continuous inspiration and for keeping the spaceship operational while he was busy fighting the onslaught of words and sentences. He also wishes to thank his daughter, Anita Day Barsukov, for nourishing his interest in explaining things and for trying to explain English to him in return.

Foreword

Jinrong and Yevgen have provided a wealth of things you didn't know you need-ed to know about using batteries in one volume. This book provides valuable insights into why batteries behave the way they do so that you can understand how to use them in an optimal fashion. The authors are recognized experts in their field and have compiled a comprehensive set of information using their extensive experience in practical battery management. As a reference work, it enables the reader to find necessary information quickly, and you will likely find not just what you are looking for, but also valuable perspectives beyond what you expected. The book is readable despite the technical depths it reaches in being so thorough. This book will certainly be perched on my desk, and I expect to refer to it often.

Andy Keates,
Battery technologist, Intel Corporation

This groundbreaking book provides, for the first time, practical approaches to battery management systems, including circuits. This book initially attracted my attention because of my great respect for Dr. Yevgen Barsukov. Early in his career, he developed a practical approach to battery modeling, and later he invented the most commercially successful fuel gauging system, Impedance Track™. I was pleasantly surprised to find that I actually enjoyed reading this book; the authors present the subject with some humor in an easily understand-able way.

This book explains both the "how" and "why" of battery management systems and provides would-be designers of battery management systems and chargers with enough information to be successful. Electrical engineers will find this book to be essential reading, or else discover (by costly mistakes) the many pitfalls in designing battery management systems. This book, by the examples it provides, makes me optimistic that portable electronic devices will continue to improve.

Robert Spotnitz
President, Battery Design LLC

"It's just a battery, how hard can it be?"

Well, think again! A small battery is a work of wonder, and today only a handful of people know how to make one that is safe and performs well.

How lucky that two of those people, Yevgen Barsukov and Jinrong Qian, should be so generous as to share their deep understanding of small Li-ion batteries with us! Look inside just about any consumer battery, and you're likely to find a Texas Instruments chip. Well, guess whose genius is behind that chip? Yes, the authors of this book invented the technology in that chip! It is the holy grail of batteries because with it one can reliably tell know how much charge is left in the battery.

While I have admired Yevgen's technology for years and attended fascinating talks by him and Qian, after reading this book, I am positively in awe of them. The book taught me many interesting and essential details of batteries and Li-ion cells, such as how to prevent people from using poorly-made knock-off batteries. The book reads so easily and is so full of practical advice, guiding you in your component selection and in the prevention of glitches that could cost your company millions.

Finally, the full gamut of Li-ion batteries is fully covered by this book (which focuses on small Li-ion batteries), and by my book (which focuses on large batteries instead).

Yes, it's just a battery, and it *will* be easy... after you have enjoyed reading this book from cover to cover!

Davide Andrea,
author of "Battery Management Systems
for Large Lithium-Ion Battery Packs"

1

Battery Chemistry Fundamentals and Characteristics

1.1   Introduction

It is tempting to think about a battery as just a constant voltage source, basically a power supply that you set to a given voltage and forget about. A battery, however, is a little portable chemical factory with limited available resources. Because of its chemical nature, it has a complicated voltage response to an applied load that can sometimes affect your device performance and a limited run time that depends on the type of the load you have in the system. This response also strongly depends on temperature and battery age.

To provide a solid foundation for your understanding of battery-powered portable systems, this chapter describes in detail all considerations that should be reviewed when selecting a battery for a specific application. Then we go from the general to the specific by providing details on the characteristics, strengths, and weaknesses of each of the popular battery chemistries.

This chapter then summarizes the best handling practices based on a particular chemistry for the most common rechargeable battery types in order to ensure the best longevity of the battery as well as adequate safety of the device. This discussion includes requirements regarding the electronics that service the battery, namely, the charger and safety devices, that are also specific for each chemistry. Finally, we go over possible developments for each of the battery chemistries to give you an idea about what to expect in the future.

1.2   Battery Fundamentals and Electrical Behavior Under DC and Transient Conditions

Portable batteries share the same general principles as primary batteries and other electrochemical power sources, such as fuel cells. They consist of two electrodes that are electrically connected to active materials and immersed in an electrolyte with a porous separator placed between them to prevent electric contact, but allow ionic flow. Figure 1.1 shows a schematic for a common battery arrangement.

The *positive (+) terminal* of a battery is connected to an electrode covered with an aggressive oxidizing material that is capable of ripping electrons from other materials. One example of a common oxidizing material is oxygen (which is indeed used in the cathodes of fuel cells), but in batteries solid oxidizing materials such as MnO_2 and $NiO(OH)_2$ are used. Such an electrode is called a *cathode* in batteries, and materials used are called *cathode materials*. Use of this naming convention rather than the sometimes used term *positive electrode* makes it clear that neither the electrode nor the material used by itself has any electric charge unless assembled in a battery. The *negative (–) terminal* of a battery is connected to an electrode covered with a strong reducing material that is rich in lightly bound electrons and can easily give them away. These materials are similar to "fuel" in its function. Indeed some common fuels, such as natural gas, can be used as an anode agent in fuel cells. In batteries, it is more practical to use solid fuels, such as Cd or Zn metals, or more exotic lithium intercalated into graphite. Such an electrode is called an *anode* in batteries, and its materials are called *anode materials*.

Again, note that the words *cathode* and *anode* in electrochemistry (and in some industrial applications, such as metal electrodeposition) are used differetnly because the electrodes themselves are neutral and do have neither oxidizing nor reducing agents permanently attached to them. Current direction can be chosen by experiment since direction is determined by an external power source. A reversal of current by an external power source can cause either an oxidizing or reducing reaction on the working electrode. If electrons are passed

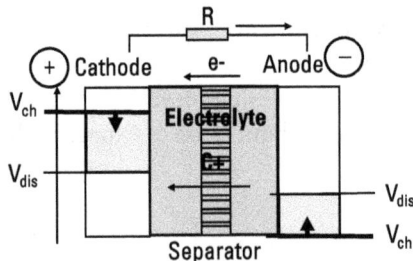

Figure 1.1   Schematic of an electrochemical power source.

to the electrode, it is called a *cathode*, when they are forced out of the electrode, it is called an *anode*.

In contrast, in batteries the oxidizer and fuel are attached to the electrodes permanently. Because the purpose of the battery is to be a power source, the naming of electrodes is based on their open circuit voltage, which defines the electron flow direction when the battery is acting as a power source, for example, when electrons are flowing spontaneously. In such a case electrons are going to the oxidizing electrode and out of the fuel-covered electrode. By analogy with electrochemistry (but only for the case of spontaneous electron flow), in batteries the electrode connected to the positive terminal and receiving electrons is permanently called the *cathode* and that connected to the negative terminal is called an *anode*, even if the battery is rechargeable and current direction can be reversed by the external force of a charger.

The key to battery operation is to harness the "desire" to pass electrons from anode to cathode material that exists in the system because, given overall electric neutrality, the combination of electrons and cathode is more thermodynamically stable than the combination of electrons and anode. All chemical systems are changing in the direction of larger thermodynamic stability, similar to how water flows down to minimize its gravitational potential energy. While we are not allowing these materials to react directly, we give the electrons an external path to flow when we connect the load. It is somewhat similar to building a dam, which prevents a river from flowing downward, but allows it to pass through a narrow passage where we can convert the energy of the flow into useful work.

If the oxidizing and reducing agents were to come into direct contact, they would react with each other releasing energy in the form of heat. This is not desirable in a battery and, in fact, could create a fire or even an explosion in the case of energy-dense batteries, and of course energy could not be used to power an external load. That is why active materials are kept apart by a *separator*. The separator is one of the more high-tech and complicated parts of a battery, because it has to fulfill conflicting needs. On one hand, it has to be very mechanically strong, so it will keep active materials apart even in the case of mechanical or thermal damage to the battery. It also has to have very small pores so that any growth of a metal on the anode (which sometimes happens during charging) does not allow the newly deposited metal to touch the opposite electrode.

On the other hand, the separator has to be highly transparent so that ions can flow between electrodes in the direction opposite to electron flow to keep overall electric neutrality; otherwise, too much internal resistance will cause a large power loss. One reason why a "lemon" battery can light up an LED only very dimly even if several of them are connected in series to provide enough voltage is that lemon is not a very good ionic conductor so a lemon battery has huge internal resistance. Note that sticking the metal wires close to each other into the lemon can make quite a bit of a difference in the brightness.

Both rechargeable and primary batteries have a fixed amount of active materials, which are physically attached to the electrodes. That is why every battery has a maximal discharge capacity that corresponds to complete conversion of active materials into their discharged (most stable) state, which is one of the most important battery characteristics. The active materials of fuel cells are either gas or liquid; therefore, more material can be added to replace the used material.

Use of a large amount of active material also increases the surface area for the charge transfer reaction between active material and electrolyte. For this reason the larger the capacity of the battery, the lower its internal resistance. This is one difference between batteries and fuel cells—since bulk fuel is not by itself participating in the charge transfer reaction until it is brought into contact with electrodes, a larger amount of fuel does not translate into a larger power capability for fuel cells as it does for batteries. Only the electrodes' active area determines a fuel cell's power capability. This is why fuel cells usually struggle to provide enough power (and are often used with batteries as power backup), while batteries are more likely to have issues with providing enough energy.

One type of rechargeable battery, called a *flow battery*, has its active material dissolved in electrolyte, so more active material can be added when it is depleted by providing fresh active material solution. These batteries are similar to fuel cells in the sense that their capacity depends only on the size of the tank, so adding more capacity has a lower cost, but the high up-front costs are determined by an electrode's size, which has a surface area defined by the needed power capability of the battery. Due to low weight ratio of active materials in the solution, such batteries have low energy density, which makes them more suitable to stationary applications.

The material in primary batteries either changes its crystalline structure or becomes electrically disconnected from electrodes during discharge, therefore making recharge impossible. Rechargeable batteries can either regrow their anode with minimal changes to the structure after recharge (as in the metal-anode case, which includes popular lead-acid, NiCd, and NiZn batteries) or keep its crystal structure completely unchanged (as in the case of intercalation batteries, such as the popular Li-ion batteries, which insert/eject ions from the crystal during recharge). In the first case degradation happens due to gradual changes in the anode morphology that eventually make the resulting metal electrode particles disconnected. In the second case, however, degradation is mostly due to parasitic surface reactions between a very aggressive active material and electrolyte. These reactions cause a growth of insulating layer that both increases charge transfer resistance and also electrically disconnects active particles from the current collector.

From an electrical point of view, batteries are often represented either as just a voltage source or as a voltage source connected in series with a resistor representing the internal resistance of the battery. Constant voltage presentation is accurate only for given state of charge of the battery with zero current. When a battery is charged or discharged, its open circuit voltage changes, as is the case with a capacitor, and does not stay constant like a voltage source. Therefore, it can be electrically represented as a capacitor, and although the capacitor description will be correct for short discharge durations (less than 1,000 sec), the capacitance will gradually change with the state of charge. Therefore, the modeling of a complete discharge would require a variable capacitance. Additional explanations are given in Section 3.2.

When current is flowing in or out of a battery, battery terminal voltage will be given as $V = V_{ocv} - I * R$ where R is internal resistance and V_{ocv} is the open circuit voltage at a given state of charge. However, this presentation is only accurate after current has been flowing for a very long time (typically more than 1 hr). The voltage response of a battery is compared to that of a capacitor in Figure 1.2.

As can be seen, a linear decrease in voltage is observed for a capacitor when the discharge current is passed. If a serial resistor is added to the capacitor, voltage immediately drops by the IR value after current onset, and return to a "no resistor" voltage level after current is removed. Battery voltage response to current step is delayed but after some time it approaches the behavior of a capacitor with a serial resistor. After termination of current, the battery voltage does not immediately return to the "no current" state, but slowly increases until

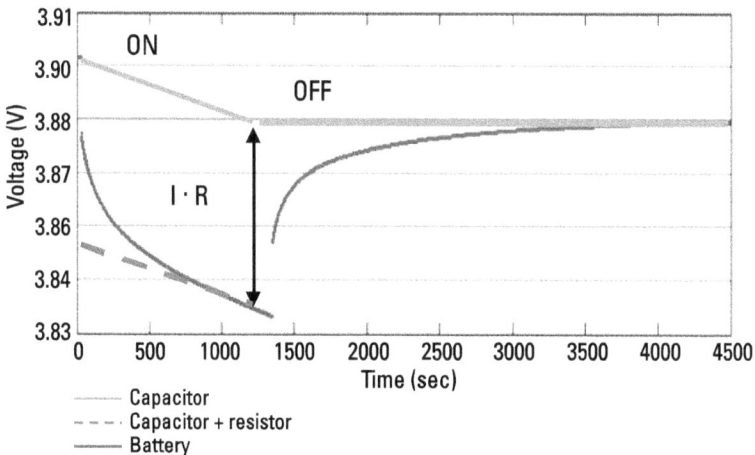

Figure 1.2 Voltage response to a step-load current applied to a capacitor, capacitor with a series resistor, and a battery.

eventually it reaches the level of equivalent capacitor voltage, which is the open circuit voltage.

What is the reason for this delay? On one hand, it is because the internal resistance of the battery is not what we are used to as resistance in electronics. It is not even completely due to the motion of electrons. We mentioned earlier two contributions to this resistance: ion flow through the separator, and charge exchange between the active material and electrolyte. But that is not all. This resistance also includes ion diffusion inside the solid particles of active material; electron conduction through the porous layer of the material as well as ionic conduction of electrolyte in the narrow pores. Figure 1.3 helps to visualize all of these charge transfer steps in an active material. It is based on a Li-ion battery, but other battery chemistries have similar overall picture.

All of these steps include not only charge conduction but also charge accumulation. You can think about them as having both resistive and capacitive components. For the same reason, a battery is also not a simple capacitor because the electron transfer between materials has to go through many different electrical and chemical stages, and each of these stages has both resistive and capacitive components.

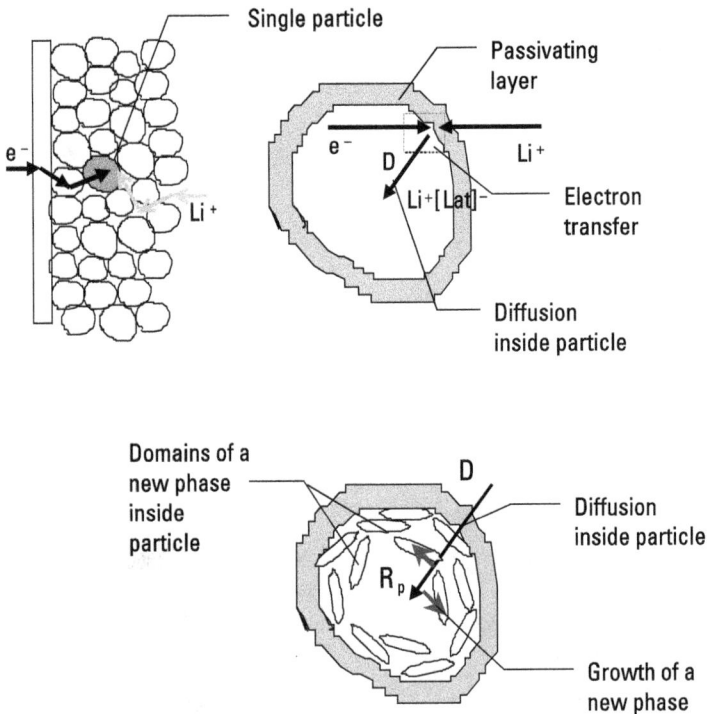

Figure 1.3   Kinetic steps involved in charge transfer in the active material of a Li-ion battery.

When considering the overall internal resistance of the cell, we need to take into account both the anode and cathode active materials kinetics steps shown in Figure 1.3, as well as electrolyte, separator, and contacts outside of the cell.

Each stage of charge transfer is associated with its own time constants, which cause complex electrical behavior. To represent battery transient behavior correctly, we should use an equivalent circuit rather than simple resistance. The simple circuit in Figure 1.4(b) is given as an example and is valid for time constants between 1 Hz and 1 mHz. For higher frequencies additional inductive and capacitive elements need to be added. Different battery types need to use different equivalent circuits if very accurate representation is required, although generic equivalent circuits allow a good enough approximation for the purposes of battery management. For the purpose of battery materials research and development where a physically meaningful description of battery kinetics is needed, even equivalent circuit presentation is not accurate enough because diffusion processes would require an infinite number of discrete elements. For more details on battery impedance modeling, see [1].

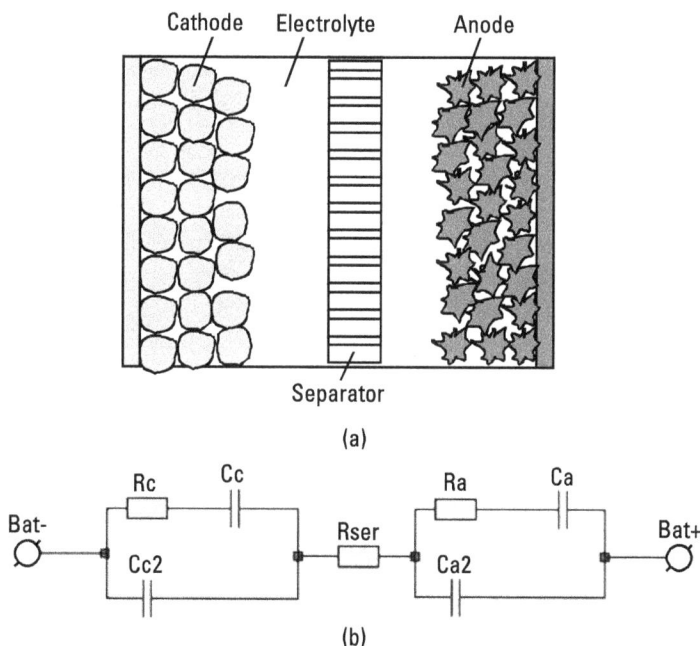

Figure 1.4 Presentation of battery conduction under variable load conditions. (a) Schematic presentation of battery components; (b) equivalent circuit describing battery conduction. Here R_c and R_a are summary diffusion, conduction, and charge transfer resistances for a cathode and anode, respectively; C_c and C_a are chemical charge storage capacitances; C_{c2} and C_{a2} are surface capacitances; and R_{ser} is serial resistance, which includes electrolyte, current collectors, and wire resistances.

1.3   General Battery Characteristics

How do we evaluate a battery for use in a particular application? Several characteristics that are applicable to all batteries and allow for comparisons to be made are reviewed in this section. They include chemical capacity, battery impedance, usable capacity at given rate of discharge and temperature, power capability, storage degradation and cycling degradation, and self-discharge rate. Monitoring and safe operating conditions are then discussed in Section 1.4.

1.3.1   Chemical Capacity and Energy

The main question regarding battery functionality in a portable application is "How long is it going to last?" This is determined by the amount of active materials, their specific capacity, and their voltage characteristics. When a battery is discharged, its voltage will gradually decrease until it reaches the minimal voltage acceptable for the device; this is called the *end of discharge voltage* (EDV), or the voltage where continuing discharge will cause damage to the battery. Integrating the passed charge during the discharge process allows us to measure the capacity Q_{max} that can be discharged until EDV is reached. Figure 1.5 shows the voltage profile during a low-rate $LiCoO_2$-based lithium ion battery discharge.

To compare batteries of different shapes and sizes, it makes sense to represent the chemical capacity in relation to battery weight (in Ah/kg) or in relation to volume (in Ah/l). Both representations have their benefits depending on what is more important for a particular application. For example, for cell phones and tablets the most admired property is a thin profile. From this point

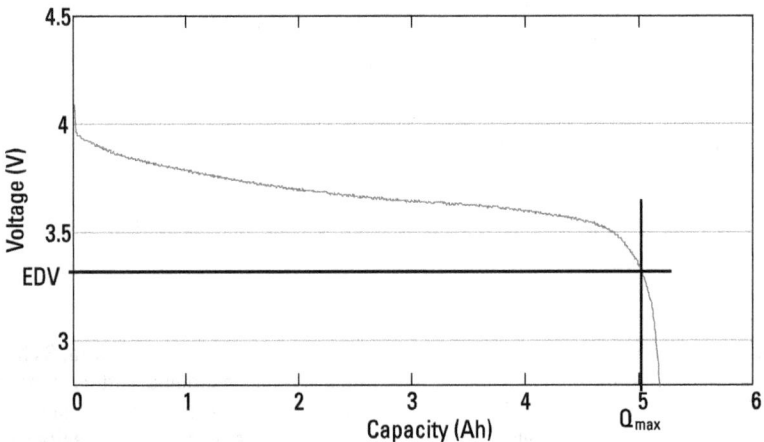

Figure 1.5   Voltage profile during low-rate discharge of battery.

of view, capacity by volume is the most critical factor, which made the use of polymer cells popular.

Because different chemistries have different voltage profiles, it is possible to have more mAh, but less energy because voltage might be lower. From this point it is useful to talk about battery energy (integral of battery voltage multiplied by current over the discharge period), expressed as Wh/kg and Wh/L, especially when comparing different chemistries.

1.3.2 Battery Impedance

It is tempting to describe the voltage/current relationship of a battery as similar to that of a resistor, for example, to assume that it will have a constant voltage drop across the cell $dV = I * R$ when current is applied. Despite the multiple kinetic steps that charge has to undergo as it travels through a battery, this approximation does work for the case of a long time discharge, when all of the transient processes have finished and steady-state gradients of concentration have been reached inside the cell. How long does it take before we can measure such a "stationary" battery resistance? To answer this question, we need to look at the full picture, that is, at the whole impedance spectrum of the battery without any shortcuts. An example of such a spectrum is given in Figure 1.6.

The impedance spectrum is represented here as a Nyquist plot, where the *X*-axis is the real part of impedance, and the *Y*-axis is the negative imaginary part. Each point on the plot corresponds to one frequency at which impedance was measured. In the case of batteries, higher frequencies will be on the left side, and the real part of impedance increases as frequencies move toward lower values to the right. This spectrum includes the effects of all discharge rate limiting factors that have been discussed earlier. Since frequency is the inverse of relaxation time, we plotted relaxation times as a second plot, which allows use to see more clearly how long you have to apply a load for a particular effect to appear. All effects that slow down charge transfer (e.g., resistive effects) will increase the real part of the impedance. All effects that increase the capacitance will affect the imaginary part of the impedance. Inductive effects do not occur in electrochemical systems, although some inductance of wires can appear at frequencies higher than 100 kHz (not shown here).

You can think about various effects as being "added" as you move from left to right, from shorter load application times to longer. I invite you to take a short trip along the impedance spectrum with us using Figure 1.6:

1. We start at the highest frequency (shortest relaxation time, 0.1 ms) and only see the effect of metallic current collectors. Since they are purely resistive and have negligible capacitance, you only have the real part of the impedance, and the imaginary part is zero. This means that

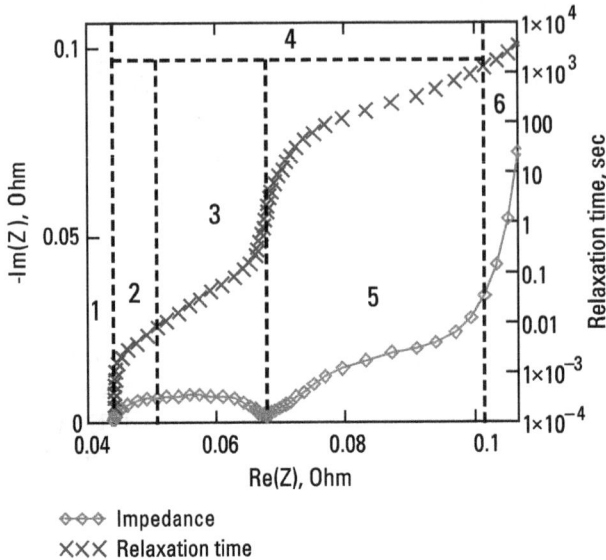

Figure 1.6　Impedance spectrum of a Li-ion battery represented as a Nyquist plot with the X-axis showing the real part of impedance and the Y-axis the negative imaginary part. Diamonds show impedance points, and crosses show relaxation times that correspond to each impedance point along the X-axis. Area 1 includes ohmic resistance of current collectors and separators; 2—resistance and capacitance of solid-electrolyte interphase (SEI) on particle surface; 3—charge transfer resistance and double-layer capacitance; 4—distributed resistance of active material and ionic resistance of electrolyte in pores; 5—diffusion effects; and 6—saturation of total resistance and storage capacitance due to limited length diffusion toward the bulk of material particles.

if we only apply a 1-ms load pulse to the battery, the voltage response will be exactly like that of a resistor of about 40 mΩ.

2. Then we proceed to the right to the longer relaxation times above 1 ms and observe an increase in both the real and imaginary parts. This increase is due, first, to the passivating layer on the surface of the particles, known as the solid-electrolyte interphase (SEI) because it provides both protection from a direct reaction with the electrolyte and the ionic conductivity needed for the charge transfer. It has both a geometric capacitance like any thin insulator surrounded by conductors, and resistance to ion transfer through the layer. These effects together can be thought of as a resistor and capacitor in parallel and they create a semicircle on the Nyquist plot.

3. At even longer relaxation times (i.e., starting at 10 ms), the effect of charge transfer resistance and double-layer capacitance of the electrolyte at the boundary with the electrode creates another semicircle, usu-

ally with a larger diameter. Note that the semicircles from effects 2 and 3 can overlap and appear as one elongated semicircle.

4. At the same frequencies as 2, 3, and 5, the effect of the distributed resistance of the active material and electrolyte in pores also adds to the real part of the impedance, further stretching both semicircles 2 and 3 and increasing the slope of the line of diffusion effects in 5. This complicates our ability to determine exact values for resistances manually and requires actual fitting of the impedance spectrum to a physical function to distinguish between all of these factors. You can just keep in mind that aging of the battery will cause the resistance of the active layer to increase such that the semicircles are going to stretch more and more, increasing the total steady-state impedance of the cell. By the time we have added all effects from 2, 3, and 4, we get an effective resistance of 67 mΩ. This is, of course, highly variable among different cell types and even among cells in the same batch.

5. At relaxation times below 1 sec, all surface effects are finished, and diffusion effects take over, further increasing both the real and imaginary parts of impedance equally. This produces a characteristic for diffusion (as well as for transmission lines in electronics) 45° line in the Nyquist plot. Some deviations from the 45° line usually happen as we proceed to even longer time constants due to nonhomogeneous particle sizes and shapes that cause a distribution of diffusion length.

6. Finally at time constants longer than 1,000 sec, effective resistance (in this case, the real part of impedance) stops increasing. This final resistance is what will correspond to the IR drop of the battery when a steady load is applied, and we can talk about it as a the *DC resistance* or internal resistance of the battery. In this case, it is about 108 mΩ, that is, 2.7 times larger than resistance measured at 1 kHz. Note that cell makers often report cell impedance at 1 kHz for historical reasons and also due to simplicity of the measurement. But as you can see it is much lower than the DC impedance you will observe in a continuous discharge. Also when battery impedance aging is reported at 1 kHz (1-ms relaxation time), it is misleading because it does not include most of the actual changes in battery materials that only appear at time-constant ranges 2, 3, and 4 and do affect final DC resistance.

At the same long relaxation times above 1,000 sec, the negative imaginary part of impedance starts increasing with increasing relaxation times as if it is a serially connected capacitor. Capacitance that can be measured from this dependency is huge (thousands of farads) and corresponds to actual chemical storage of the battery in the bulk of solid particles and not to some kind of surface effects. A combination of saturated resistor value and storage capacitor gives an equivalent circuit for the battery that can be used for steady-state discharge. One thing to keep in mind is that these resistor and capacitor values depend on the state of charge of the battery, so a simple "constant values" model can only by used for short time discharges. A nonlinear model with variable R and C would have to be used to describe complete discharge.

1.3.3 Usable Capacity

Battery voltage depends not only on its state of charge, but also on discharge current. This is caused by voltage drop IR due to battery internal resistance. Note that initial drop is lower due to the transient effects described earlier, so it takes about 500 sec before full IR drop can be observed. This drop is higher at high currents, low temperatures, and for aged batteries, which have increased resistance. How does it influence run time? Minimal system voltage will be reached earlier, therefore reducing the "usable" capacity that the battery can deliver, as illustrated in Figure 1.7.

Capacity integrated until voltage reaches EDV under load conditions is called *usable capacity*. Because it depends on current and temperature, it has to be specifically evaluated for each application. Note that internal resistance R depends on the state of charge and temperature and increases at the end of discharge, as can be seen in Figure 1.8.

Therefore, simple modeling assuming fixed R will not give an accurate estimate of usable capacity. Also battery manufacturers often report battery impedance at 1 kHz. This value cannot be used as an estimate for internal resistance at DC conditions, because low frequency impedance (that corresponds to DC conditions) is much higher than that at 1 kHz. Although DC resistance is

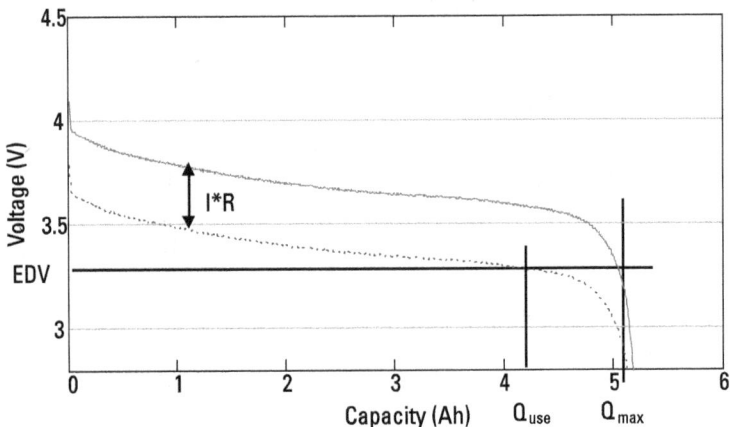

Figure 1.7 Battery voltage profiles under load (dotted line) and without load (solid line).

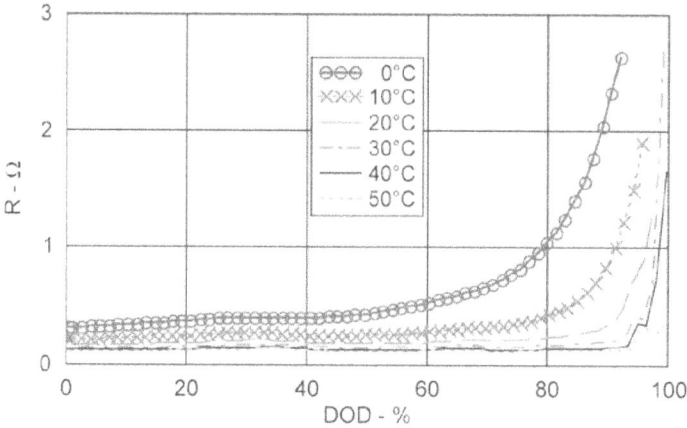

Figure 1.8 Temperature dependence of low-frequency impedance for a Li-ion battery at depth of discharge (DOD) levels.

typically two to three times the 1-kHz resistance for charged Li-ion cells, the ratio is unpredictable for different states of charge, aged cells, or at low temperatures, so it is better to always measure DC resistance directly. Impedance of 1 kHz is useful for detecting catastrophically failed cells during production because of the fast measurement ability afforded and widely available instrumentation. If low-frequency impedance data are not available for detailed modeling, two of the best ways to estimate usable capacity is to perform a test in an actual device or to refer to manufacturer discharge curves at loads comparable to the expected application loads.

Because of the strong temperature dependency of impedance, it is very important to consider the actual temperature environment of a device. In many cases battery self-heating (especially if combined with heating for the device itself) can provide a reasonable run time even at very low outside temperatures. If a device can survive without shutting down during the initial period before its battery has had a chance to warm up, it can continue operating almost as long as at room temperature, as can be seen in Figure 1.9.

If you look at the –10°C voltage plot, there is an initial sharp drop in voltage due to high impedance at low temperature, which is followed by a raise due to self-heating. Knowing this effect, it is critical to test your battery either in the device itself (which is best) or in a thermal box that has heat exchange properties similar to those of the actual device. The latter method is much less advisable because heat coming from electronic device operation is an order of magnitude higher than the heat generated by the battery itself. Some other devices that convert battery energy into mechanical work can dissipate energy outside of the device, so this rule has to be used with a good understanding of heat dissipation in each case. In any case use of a thermal box is better than testing the

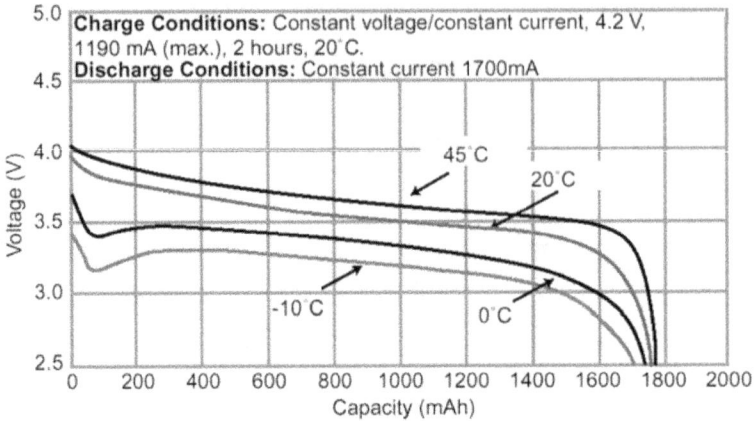

Figure 1.9   Voltage profile of Li-ion battery discharge using 1°C rate at different temperatures.

battery discharge under open-air conditions because this type of test at least provides consistent heat exchange conditions from test to test. Note that just a fan blowing at the battery in a thermal chamber increases the heat exchange rate about five times (given the same air temperature!), which can make a difference between success or failure in terms of satisfying required usable capacity.

1.3.4   Power Capability and the Ragone Plot

To evaluate how much energy can be obtained from a battery at different rates of discharge (in particular, at given power), a plot of energy density versus power density is often used. It is called a Ragone plot (see the example in Figure 1.10). This particular plot shows relative values referred to weight, but can be also useful to refer to volume or to surface area depending on the desired critical factor in a particular application.

Each battery has a "critical" power density where its usable energy starts dropping rapidly. This corresponds to currents causing a voltage drop to be high enough to cause the flat portion of the voltage curve to approach the EDV voltage level. Looking again at Figure 1.7, we can see that as long as EDV is reached in a steeply decreasing portion of the voltage curve, small changes in IR drop cause only small changes in usable capacity. However, even a very slight additional increase in the IR drop causes a large percentage drop in usable capacity when the flat portion is approached.

From this comparison we can determine that a cylindrical cell has a higher energy density at low rates than a laminated pouch cell; however, at high rates the laminated pouch cell shows superior rate capability because its critical power is almost the same as that of a cylindrical cell. Calculations of the battery

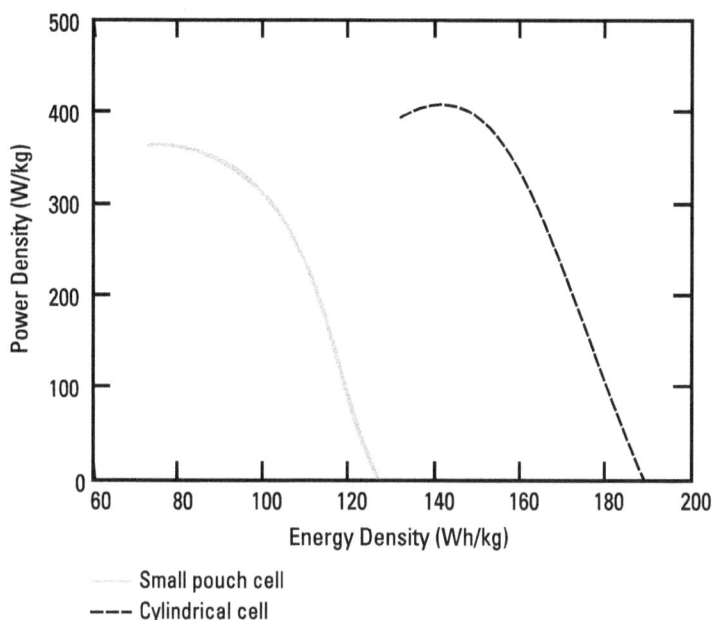

Figure 1.10   Ragone plot comparing cylindrical cell with laminate pouch Li-ion cells.

cell size needed for a particular application should ensure that the application power does not exceed one-third of the plot's height. That will guarantee that load spikes or an impedance increase with aging will not cause the battery to go over the critical threshold.

1.3.5   Durability, Cycle Life, and Shelf-Life

The capacity and rate characteristics of a new battery do not stay unchanged as the battery ages. Changes can be summarized as *loss of chemical capacity* and *impedance increase.*

Capacity loss is caused, for example, because the crystalline structure of rechargeable material will start to change. Particles can break up, became disconnected from bulk material, and therefore no longer participate in energy storage. Many other chemical degradation mechanisms cause the same overall effect of having less material available in the battery. Chemical capacity loss will equally influence high-rate and low-rate applications.

Increases in impedance are caused primarily by the growth of passivating layers on the surface of active material particles. These layers decrease the conductive path for electrons as if we had ground away the wires connecting them. Another mechanism of impedance increase is the loss of electrolyte in pores, thus decreasing ionic conductivity. An increase in battery impedance is most critical for the high-rate applications. Indeed, in applications such as a TV remote or watch, where currents are close to zero, IR drop remains negligible even as R is increasing. However, in applications with rates of discharge of C/3 up to 1C rate (C-rate means current causing full battery discharge in 1 hr; C/3 refers to discharge in 3 hr), the usable capacity loss because of an impedance increase can be the main contribution to overall loss.

How much chemical capacity loss and impedance increase can we expect? The effect is different for different batteries; commonly chemical capacity loss is much less than an impedance increase. Figure 1.11(b) shows voltage profiles for low-rate discharge during 100 cycles of a Li-ion battery. It can be seen that capacity at EDV of 3V changes only by about 5%. However, low-frequency impedance change, shown in Figure 1.11(a) for the same 100 cycles, is more than 60%. Note that change at the low-frequency part of the impedance (which is most important for DC discharge) is very large, while impedance at 1 kHz almost does not change. This highlights the importance of obtaining low-frequency impedance for evaluating battery aging. While dedicated impedance spectrometers exist for this purpose, a single point measurement of low-frequency impedance can be done, for example, by comparing open circuit voltage (OCV) before the pulse with voltage after applying load I for 100 sec, V_1. Approximate DC resistance will be $R = (V - OCV)/I$, assuming that discharge current is entered as negative.

The rate of degradation in any particular application strongly depends on usage conditions, mainly voltage (state of charge) and temperature. Sometimes the effect of the state of charge can be opposite for different battery types. For example, lead-acid batteries benefit from being stored at high states of charge, whereas Li-ion batteries prefer more mild conditions of lower voltages, as can be seen in Figure 1.12.

Although these data show the degradation during storage, a similar relationship exists for degradation during cycling, because the parasitic reaction with electrolyte will also accelerate with voltage and temperature when current is passing. To make matters worse, cycling of batteries causes expansion/contraction of particles, which causes cracking of the passivating layer (which needs to regrow as a result). Regrowth of the passivating layer to heal the cracks uses up some of the available Li and also clogs the pores with accumulating passive reaction products. The thickness of the passivating layer also increases. Shrinkage/expansion of particles also causes mechanical changes in the structure of the active layer consisting of many particles that all change volume. Such mechanical changes can cause electrical disconnect of the particles from each other and from the binder and conductive additive. Overall, cycling at a typical rate accelerates the overall impedance increase and capacity loss by about 5 to 10 times compared to the storage degradation during similar time durations. An example of capacity degradation with cycling is given in Figure 1.13.

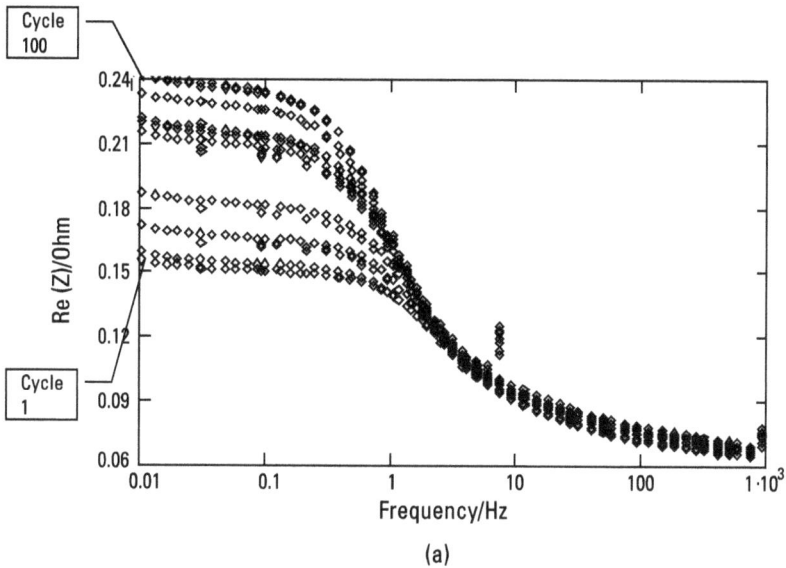

(a)

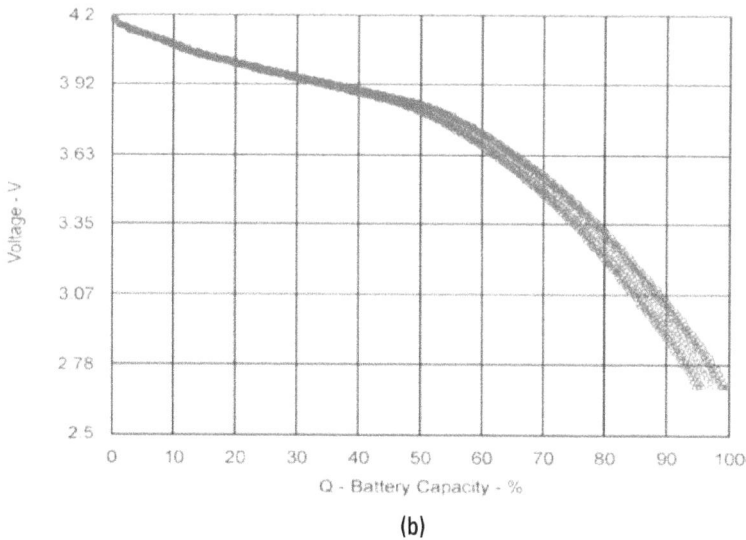

(b)

Figure 1.11 Results of a C/10 rate discharge cycling test: (a) Impedance spectra measured from 10 mHz to 1 kHz after each 10 cycles of a 100-cycle test. (b) Voltage profiles measured during each cycle.

1.3.6 Self-Discharge Properties

Batteries will not retain their charge forever. Charges can be lost via many mechanisms, depending on battery chemistry and design. Here are a few of the main mechanisms:

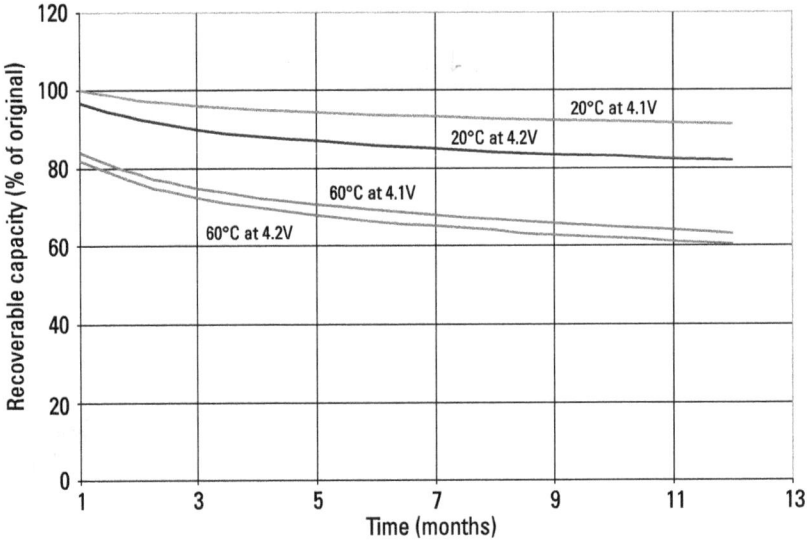

Figure 1.12 Degradation of LiCoO₂/graphite Li-ion battery capacity during storage at different voltages and temperatures. (Based on data from [2]).

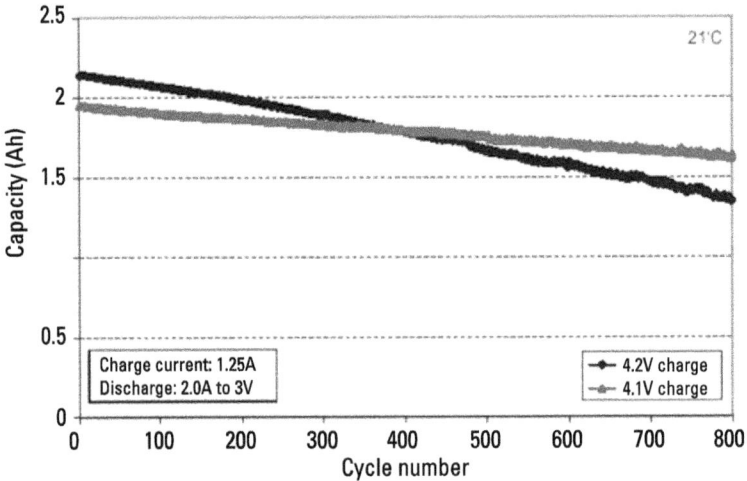

Figure 1.13 LiCoO₂/graphite based Li-ion battery capacity degradation with cycling using different maximal charging voltages. Charging to 4.2V causes higher initial capacity, but it decreases faster with cycle number.

- *Parasitic conductance inside battery.* Dendrites grown during charge on the anode or defects in the separator might short the electrodes. Nanoporous separators have been successfully used to reduce this effect, although nonwoven fabric separators remain most common. This mechanism does not apply to Li-ion batteries, in which a dendrite can be catastrophic, but rather to metal-anode batteries such as lead-acid and NiCd batteries.

• *Shuttle molecules.* Some molecules can become oxidized (lose an electron) on the cathode, diffuse to the anode, and become reduced (receive an electron, returning to original state) there and diffuse back to the cathode. The same molecule can be reused indefinitely, eventually discharging the battery. Normally such molecules are avoided during battery manufacturing and only occur as impurities, with better quality batteries having fewer such shuttles. An exception are some shuttle molecules that can be oxidized only at the high voltages that occur only during overcharge. Such molecules can provide additional safety for a Li-ion battery effectively shorting it out when maximal voltage is exceeded, so they can be added on purpose.

• *Recombination of oxygen-hydrogen.* Some hydrogen development happens on the anode (water reduced) and oxygen development on the cathode (water oxidized). Since separators in water-based batteries are not airtight, gasses then diffuse through the separators and react, forming water again and discharging the battery. This process also generates heat, which becomes especially noticeable close to the end of charge.

• *Impurities.* Impurities in electrolyte can get oxidized and reduced, discharging the anode and/or cathode independent from each other.

In Li-ion batteries, electrolyte reacts with electrodes because of their extreme activity.

In NiCd and NiMH batteries, voltage depression (memory effect) reduces battery energy without reducing discharge capacity. This happens due to changes in the crystalline structure of the cathode material to a more dense, passive form when stored for a long time in the charged state.

The rate of self-discharge increases with temperature because all reactions are accelerated and with battery age (cycles) because fractured active material has a higher (more active) surface area. Figure 1.14 gives an example of the temperature dependence of self-discharge for a NiMH battery. Note that exact rates are dependent on particular cell design.

1.4 Monitoring and Safety

Increased temperature not only accelerates degradation, it can also cause thermal runaway and battery explosions. This is a specific concern with Li-ion batteries because of their very aggressive active material. To give an example, a Li-

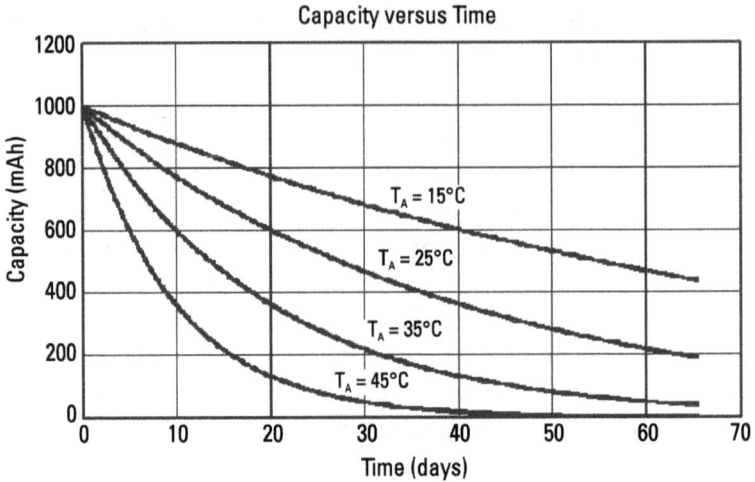

Figure 1.14 Temperature dependence of NiMH battery self-discharge.

carbon intercalated compound reacts with water and the released hydrogen can be ignited by the heat of the reaction. Both cathode and anode materials start reacting with electrolyte when the temperature goes above the thermal runaway threshold. Figure 1.15 shows the heat output for Li-ion battery cathode material depending on the temperature.

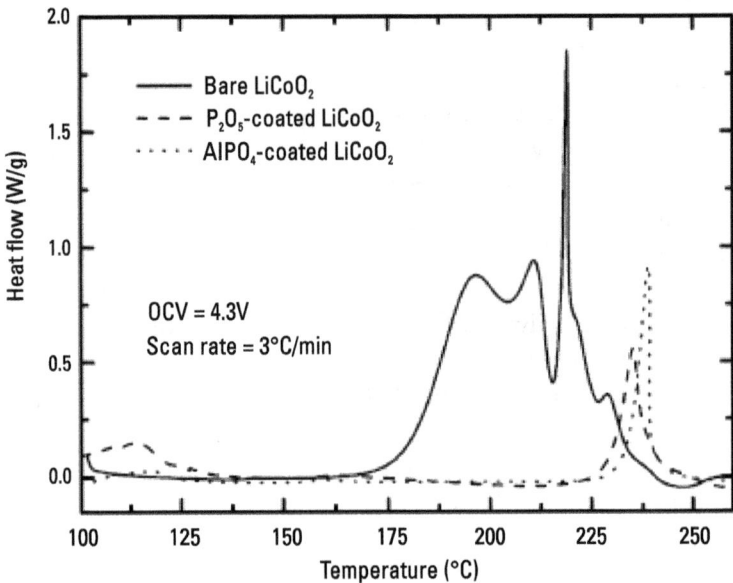

Figure 1.15 Heat flow dependence on temperature of $LiCoO_2$ charged to 4.3V. (*After:* Cho et al., *Chem. Mater.*, 15/16 [2003], 3192.)

Other batteries are also capable of explosive self-heating, just because they store a large amount of energy in a small volume. For example, NiMH batteries will start to produce significant self-heating above 80°C and, in the absence of good heat exchange, would undergo thermal runaway. A rapid temperature increase can happen if a battery is overcharged at a high current or shorted. During overcharge of a Li-ion battery, very active metallic lithium is deposited on the anode. This material highly increases the danger of explosion, because it can explosively react with a variety of materials including electrolyte and cathode materials. Figure 1.16 shows a result of such an overcharge event.

Material developments are ongoing to increase thermal runaway temperature. However, it is always going to be a responsibility of the system designer to provide for correct charging conditions as well as for the possibility of multiple failures of electronic components, which by design should still not cause catastrophic failures. Typically this means having at least two independent overvoltage and short-circuit protection circuits as part of the battery pack. We will discuss protection circuits in more detail in the Li-ion battery section.

Some manufacturing defects can significantly undermine battery safety. An internal short-circuit inside a battery cell cannot be prevented by an external protector. Such a short circuit can cause such extreme localized heating that thermal runaway conditions are reached locally and can propagate from the short-circuit spot until the entire battery is engulfed in the inferno. Therefore, it is extremely important during cell manufacturing for any foreign particles, especially metals, to be eliminated. Metallic particles inside cells have caused

Figure 1.16   Battery pack that exploded after overcharging.

several device fires and massive battery recalls. While many new methods to prevent them and detect them during manufacturing are being developed, the level of adoption of the various methods differs among manufacturers. As a device maker it would be your responsibility to verify that your chosen cell maker is following state-of-the-art procedures during manufacturing and testing to ensure the absence of metallic particles.

Sometimes a defect can develop during the battery aging process; for example, if an imbalance between the amount of cathode and anode material develops due to uneven aging. For example, delamination of an anode's active material can cause metallic Li deposition directly on the Cu current collectors. Being extremely active, dendritic Li powder is waiting to be ignited by a microshort, which could also cause a catastrophic event. From this point of view, it is very important to continue monitoring the battery as it is being used (e.g., via battery monitoring and gauging ICs) and if abnormal aging is detected, the battery should be taken out of operation before an issue occurs. This kind of battery-aging monitoring is just starting to become common and is available in the newer battery monitoring ICs.

1.5 Overview of Different Battery Technologies

1.5.1 Lead Acid

1.5.1.1 Active Materials and Reactions

Discharge

Minimum 1.5V.

The anode (lead) and cathodes (lead dioxide) are converted to lead sulfate.

The electrolyte is converted to water.

This conversion reduces the electrolyte density and this is why the density of the electrolyte can be used to determine the state of charge (SOC), for example, by using sink-float standards.

During use, the electrolyte water can be lost by evaporation as well as hydrogen/oxygen evolution and must be replaced.

Charge

Maximum 2.7V.

The negative electrode is converted to lead.

The positive electrode is converted to lead dioxide.

Sulfuric acid is added to electrolyte.

Minimum current charge termination can be used.

Trickle charges are possible.

Reactions

See Table 1.1 for lead-acid battery reactions.

1.5.1.2 Construction

A typical lead-acid battery consists of mold lead plates inserted in liquid concentrated sulfuric acid and enclosed in a robust polymer casing equipped with a vent. Depending on the starting or deep-cycle purpose of the battery, the thickness of the plates varies (deep-cycle batteries use thicker plates). Other variations typically include use of a glass-matt separator or gelating agent to prevent shorting of the plates by $PbSO_4$ dendrites. Some high-rate batteries use a wound design and thin lead foil instead of bulk plates. Figure 1.17 shows an example of a typical lead-acid battery design.

1.5.1.3 Charging Method

A lead-acid battery is very flexible in terms of charging. There are two limitations: (1) The charging rate is limited by extensive gassing and temperature. If neither is observed, the rate can be increased. Usually higher rates are acceptable at the beginning of charge. (2) Voltage is limited by 2.65V during the high-current period of charge, but should not exceed 2.39V/cell at 25°C if current is low. If these two conditions are satisfied, the design of the charger is limited more by hardware limitations (such are rate capability of power supply) than by battery limitations.

The most common charging methods are listed below:

1. *Constant current (CC)*. Current at about a 12-hr (C/12) rate is applied. Charging is terminated when the voltage exceeds 2.65V/cell.

2. *Two-step constant current*. Current at a high (C/6) rate is applied for about 3 hr, later switched to a lower (C/16) rate that continues for 8 hr. There is no voltage control in this case.

Table 1.1

Lead-Acid Battery Reactions

Positive Electrode	Negative Electrode	Electrolyte	Discharge →	Positive Electrode	Negative Electrode	Electrolyte
PbO_2	Pb	$2H_2SO_4$		$PbSO_4$	$PbSO_4$	$2H_2O$
Lead dioxide	Lead	Sulfuric acid	← Charge	Lead sulfate	Lead sulfate	Water

Example of construction

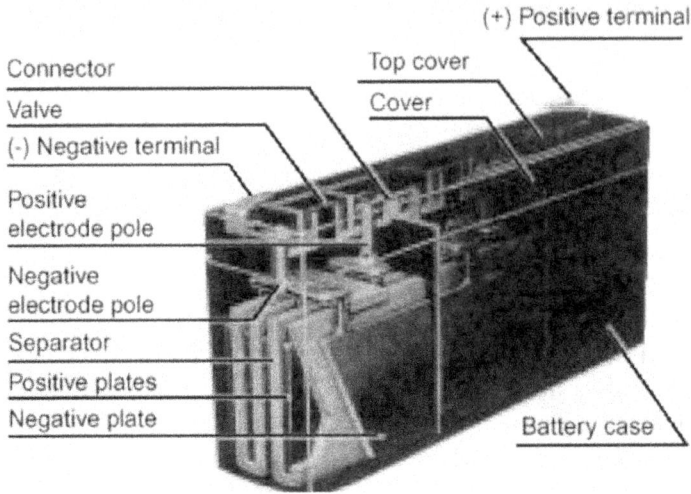

Figure 1.17   Typical lead-acid battery design.

3. *Constant current/constant voltage (CC/CV)*. Current at about a 1C rate is applied until the voltage reaches 2.39V/cell. Then it is switched to a constant voltage charging at the same voltage. Charging is terminated when the current drops below the C/20 value.

4. *Constant voltage*. Voltage of 2.39V/cell is applied to the cells directly. Current is limited by the current capability of the power supply. Charging is terminated when the current drops below the C/20 value. Note that if charging is not terminated and voltage is maintained indefinitely (as in the case of backup batteries), it is important to follow temperature compensation of the float voltage to avoid excessive hydrogen evolution, which causes electrolyte loss and plate corrosion. A compensation table (V/T) can be obtained from battery manufacturers.

Method 3 is preferable as the fastest and least harmful to the battery, but it requires more complex control. Example voltage and current curves resulting from this method, using various initial CC currents, are given in Figure 1.18.

The easiest charger design includes using an IC, such as bq2031, that is specially tailored for use with lead-acid batteries. Other ICs designed for charging Li-ion batteries can also be adapted to lead-acid charging if the CC/CV method is used and the CV voltage is set to 2.39V.

Lead-acid batteries are often used as backup batteries, so they often stay in a fully charged state. This application is well aligned with the fact that most

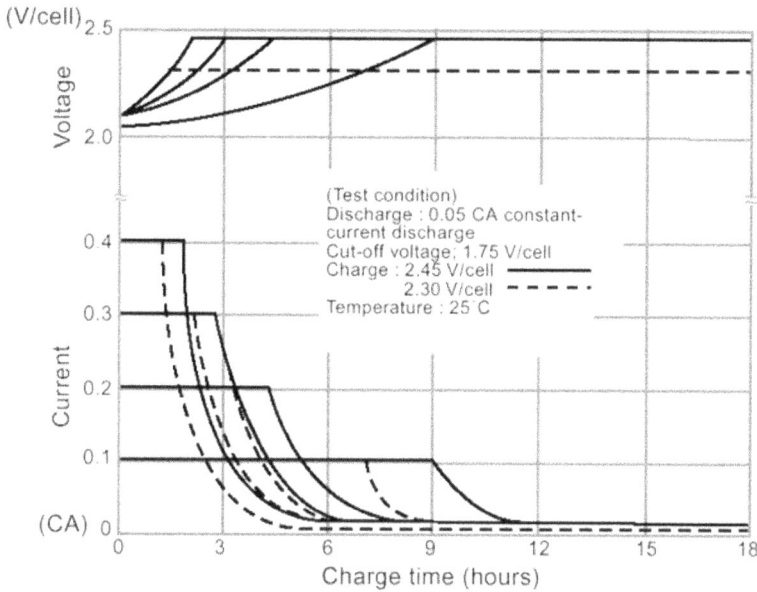

Figure 1.18 Charging profiles of a lead-acid battery using the CC/CV method with various voltage and current thresholds.

degradation mechanisms of this battery happen in the discharged state. However, care has to be taken in such applications that excessive corrosion is not caused as a result of too much charging, and at the same time that the battery is not undercharged, which would result in accumulation of lead sulfate in a highly crystallized form that cannot be converged by charging. Charger design for "always charged" operation should include periods of higher voltage to completely convert lead sulfate (charging can go up to 2.45V in this case for short period), followed by a lower "float" voltage below the 2.39V level that ensures that not too much hydrogen generation occurs, which promotes corrosion and water loss. Float voltage is somewhat temperature dependent. In variable temperature conditions chargers either have to ensure variable float voltage, or alternatively charging could consist of interchanging periods of CC/CV charging to a higher voltage, and then relaxation without any charging at all, until the voltage reaches a minimal recharge threshold that is set below the lowest float voltage in the expected range of temperatures.

1.5.1.4 Healthy Treatment

Healthy Habits

Most stable in charged state. Recharge whenever possible. However, avoid continuous charging, because hydrogen generation promotes corrosion and water loss.

High temperatures somewhat increase corrosion and water loss, but not as harmful as for Li-ion battery.

If deeply discharged, should be recharged immediately.

Starter batteries should never be deeply discharged because irreversible damage to the plates takes place.

For flooded cells, add distilled water if density of electrolyte is too high.

Degradation Mechanisms

Sulfation: Reduced form of anode ($PbSO_4$) is prone to growing large crystals, which can be reduced only very slowly and at higher voltages. Prevention—keep charged or recharge immediately after discharging. Repair—charge at elevated voltage (2.5 to 2.6V) for several days; however, only partial recovery is possible for severe cases.

Plate corrosion: Prevention—keep charged, but do not exceed temperature-adjusted float voltage (2.4V at room temperature). Repair—not possible.

Water loss: Prevention—do not overheat. Repair—is sealed, not possible. Add distilled water in flooded cells.

Usual Applications

High power, stand-by, backup systems.

Automotive starter batteries.

Deep-cycle batteries for fleet electric forklifts, cars, golf carts, or marine applications that require only limited range.

Backup power for telecommunication and computer servers. Popular because can be trickle charged or pulse/relax charged for long time periods without degradation.

Development Trends

Despite being the oldest battery, it still being further developed due to unparalleled low cost, recyclability, and good longevity in fully charged state.

Relatively Recent Developments

Sealed valve-regulated (SVRL) batteries reduce water loss, allow maintenance-free operation.

Found in most cars.

Glass matt separator batteries prevent lead sulfate crystals from shorting the electrodes, improving longevity and allowing for closer spacing of electrodes, which improves volumetric energy density.

Gelled electrolyte (adding gelling agent like silicate to sulfuric acid) reduces water loss, allows nonvertical placement of electrodes.

In combination with the above improvements, using a wound cell design increases power density by weight due to high surface area of electrodes.

Futuristic Developments

Graphite foam electrodes will improve contact between electrodes and active materials due to high surface area. This way, sulfation and electrode corrosion are drastically reduced (increase in longevity), while energy and power densities are increased. Problems with brittleness and high cost of graphite foam have to be resolved. Firefly Energy company is working on this technology.

1.5.2   Nickel Cadmium

1.5.2.1   Active Materials and Reactions

Discharge

Minimum 0.7V.

The positive electrode is converted from nickel oxyhydroxide to nickel hydroxide.

The negative electrode is converted from cadmium to cadmium hydroxide.

The electrolyte is a water solution of potassium hydroxide.

The water in the electrolyte is converted to hydroxide during discharge.

The potassium hydroxide does not react and is added for conductivity only.

Charge

Maximum 1.6V.

The reverse reaction occurs during charge.

Toward the end of charge, the reaction coulombic efficiency decreases.

Overcharge produces oxygen and hydrogen, which then produces heat as they recombine.

Charge termination is done when voltage starts to decrease due to lower IR raise at higher temperatures. Detection of $-dV \sim 10$ mV, $dT/dt \sim 1°C/$ min can be used. At low rate or higher ambient temperatures, no $-dV/dt$ will occur, therefore termination by timer, temperature delta, or coulomb count is needed.

Reactions

See Table 1.2 for nickel-cadmium battery reactions.

1.5.2.2   Construction

The design of early NiCds was similar to that of lead-acid batteries. This design is still used for large size batteries used for power backup. NiCd batteries used in portable electronics have a wound design, where NiOOH mixed with carbon black for higher conductivity is pressed on Ni-mesh (cathode), and Cd-powder mixed with some amount of $Cd(OH)_2$ is pressed on Ni-mesh making anode. Both electrodes are wound together with a separator (polyethylene nonwoven fabric) and enclosed in a steel casing with a vent to release high pressure. A typical design of a portable NiCd battery is shown in Figure 1.19.

1.5.2.3   Charging Method

NiCd batteries can only be charged with the constant current method. The reason why a constant voltage method is not acceptable is because of a peculiar voltage decrease near the fully charged state. If the CV method could be used, charging would never terminate because current near a fully charged state would not decrease but increase. The charging profiles for a NiCd battery at different rates are shown in Figure 1.20.

The charging rate can be from C/10 to 1C. As can be seen in Figure 1.20, the voltage starts to decrease at the end of discharge. When a voltage decrease is detected by observing a negative dV/dt, charging is terminated. Note that termination should be disqualified for the beginning of charge, because during the first five minutes of charge a negative dV/dt is possible due to an impedance decrease as some insulating layers break up or due to self-heating. With a lower than C/10 charging rate, termination should be done using the timer set

Table 1.2
Nickel-Cadmium Battery Reactions

Positive Electrode	Negative Electrode	Electrolyte	Discharge →	Positive Electrode	Negative Electrode	Electrolyte
2NiOOH	Cd	$2H_2O$, NaOH		$2Ni(OH)_2$	$Cd(OH)_2$	$2H_2O$, NaOH
Nickel oxy-hydroxide	Cadmium	Water, potassium hydroxide	← Charge	Lead sulfate	Lead sulfate	Water, potassium hydroxide

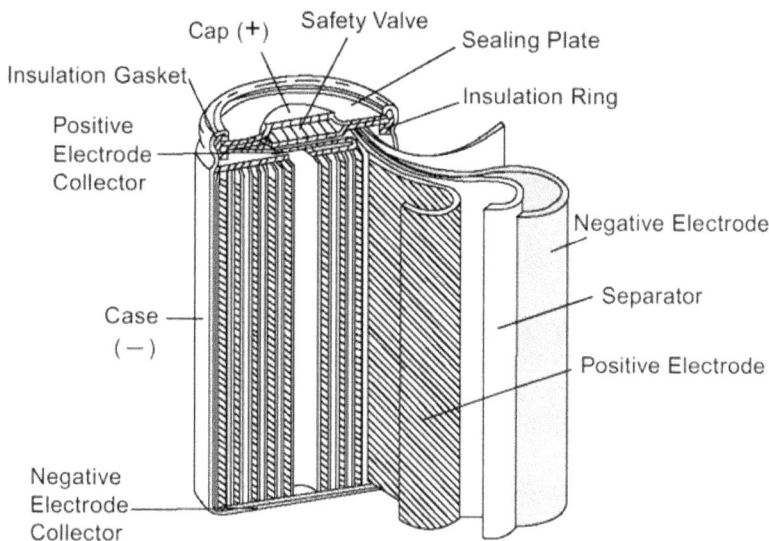

Figure 1.19 Typical design of NiCd and NiMH batteries.

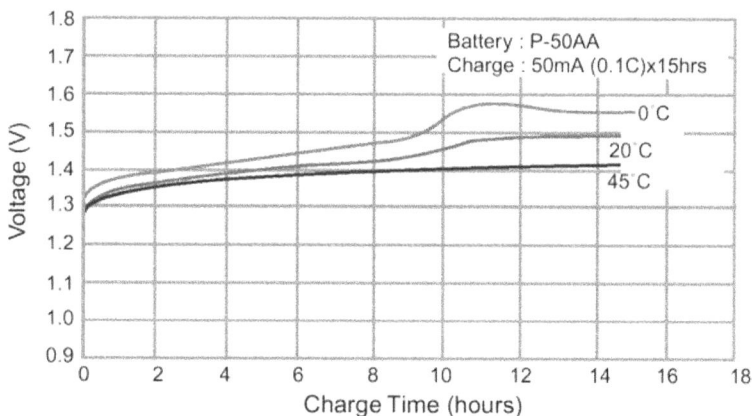

Figure 1.20 Charging profiles for a NiCd battery.

to 12 to 14 hr, because a voltage drop near the fully charged state becomes less expressed at a low rate and also because overcharge at a low rate is less dangerous. Alternatively, a temperature rise at the end of discharge can be used for terminating charging.

1.5.2.4 Rate Capability

NiCd batteries have excellent discharge rate capabilities exceeding the 10C rate and rivaled only by silver-zinc and most recently high-rate Li-ion battery chemistries like LiFePO$_4$. An example of the voltage response to different discharge rates is given in Figure 1.21. It can be seen that at a C/3 rate the battery delivers about 70% of rated capacity.

1.5.2.5 Healthy Treatment

Healthy Habits

Most stable in discharged state. However, due to fast self-discharge, to prevent overdischarge, should be placed for storage in a charged state.

If used in an application needing prolonged trickle charging (stand-by), should be periodically fully discharged, because continuously fully charged state causes a gradual change of the crystalline structure of the cathode material into inactive form.

Degradation Mechanisms

Plate corrosion: Prevention—avoid high temperatures. Repair—not possible. Very slow. NiCd has best cycle ability of all rechargeable batteries using metallic anodes with more than 2,000 cycles. However, new intercalation-type batteries like LiFePO$_4$ or those that are Li-titanate-based have exceeded this range.

Voltage depression: Cathode material changes into inactive crystalline form when stored in charged state. Prevention—discharge regularly. Repair—charge/discharge several times to 0.7V. For cases of severe capacity loss, keep at 0.7V for prolonged time (24 hr).

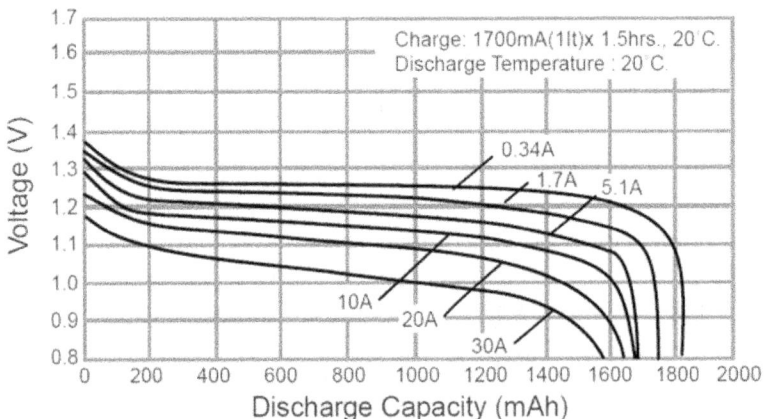

Figure 1.21   Discharge profiles of Panasonic NiCd batteries at discharge rates ranging from 0.5 to 17C.

Usual Applications

Most commonly found where following attributes are needed: high power, long usage, low cost.

Includes use in power tools, electric bicycles, solar lights, electric toys, and some use in consumer electronics.

Development Trends

NiCd batteries have become marginalized recently due to the twice or more higher energy density provided by NiMH batteries with a similar cost. At the same time concerns about toxic cadmium and difficulty recycling due to their complex cell design have caused regulatory authorities in Europe and Asian countries to encourage complete phase-out of this battery chemistry.

For the above reasons, the only new developments recently have been a side effect of active work in NiMH battery improvement. Because NiCd and NiMH share the same cathode material—nickel oxyhydroxide—improvements in NiMH have benefited NiCd. Most notably, use of spherical agglomerates of nanoparticles of the material have drastically reduced memory effect and self-discharge rates for both chemistries.

1.5.3 Nickel Metal-Hydride

1.5.3.1 Active Materials and Reactions

Discharge

Minimal 0.7V.

The positive electrode is converted from nickel oxyhydroxide to nickel hydroxide.

The negative electrode is converted from a metal hydride to metal plus water.

The electrolyte is a weak solution of potassium and lithium hydroxide.

The potassium does not react.

Charge

Maximal 1.6V.

The reverse reaction occurs during charge.

When the metal is converted to a metal hydride, heat is produced. Toward the end of charge, reaction efficiency decreases.

Overcharge produces oxygen and hydrogen, which then produces heat as they recombine.

Charge termination is done when voltage starts to decrease due to a lower IR raise at higher temperatures. Detection of $-dV \sim 10$ mV, $dT/dt \sim 1°C/$ min can be used. At low rate or higher ambient temperatures, no $-dV/dt$ will occur, therefore termination by timer, temperature delta, or capacity count is needed.

Reactions

See Table 1.3 for nickel metal-hydride (NiMH) battery reactions.

Here AB_5 is a hydrogen-adsorbing alloy, where A is a mixture of rare earth metals (called mischmetal) that includes lanthanum, cerium, neodymium, and praseodymium, while B can include nickel, cobalt, manganese, and aluminium.

1.5.3.2   Construction

Construction of the sealed NiMH batteries used in portable electronics is similar to that of NiCd batteries, except for the material of the anode, where a hydrogen-adsorbing alloy is used.

1.5.3.3   Charging Method

Charging of a NiMH battery is the same as for a NiCd battery. In calculating the charging current value, it is necessary to consider that for the same size, a NiMH battery has about twice the capacity of a NiCd battery. Because of that, chargers designed for AA (or other fixed format) NiCd batteries are not suitable for NiMH batteries because they would undercharge them or charging would take too long (because the charge current is too low).

1.5.3.4   Rate Capability

NiMH batteries can be typically discharged at up to 2C rate. They have higher internal resistance compared to NiCd batteries, but the rate capabilities of par-

Table 1.3
Nickel Metal-Hydride Battery Reactions

Positive Electrode	Negative Electrode	Electrolyte	Discharge →	Positive Electrode	Negative Electrode	Electrolyte
$2NiOOH$	$AB_5(H_2)_x$	H_2O, NaOH		$2Ni(OH)_2$	$AB_5 + xH_2O$	H_2O, NaOH
Nickel oxy-hydroxide	Hydrogen-adsorbing alloy saturated with hydrogen	Water, potassium hydroxide	← Charge	Nickel hydroxide	Hydrogen adsorbing alloy and water	Water, potassium hydroxide

ticular models can vary. Models specially developed for high current capability are used in hybrid electric car batteries, where NiMH batteries became the battery of choice. Rate capability at low temperatures is somewhat degraded because of a slower hydrogen reaction on the anode surface. Use of catalysts can accelerate this reaction in particular models. If low-temperature performance is important for a particular application, manufacturer discharge curves at different temperatures should be checked.

1.5.3.5 Healthy Treatment

Healthy Habits

Most stable in discharged state.

If used in an application needing prolonged trickle charging (standby), should be periodically fully discharged because cathode material changes into inactive form if kept for long in charged state.

Avoid high temperatures because corrosion accelerates and charge efficiency drops.

Degradation Mechanisms

Corrosion and fracturing of hydrogen-adsorbing alloy: Prevention—avoid long time overcharge and high temperatures. Repair—not possible. Degradation is usually faster than in NiCd case.

Voltage depression: Cathode material changes into inactive crystalline form when stored in charged state. Less expressed compared to NiCd because of newer technologies used in cathode material and electrolyte preparation. Prevention—discharge regularly. Repair—charge/discharge several times to 0.7V. For cases of severe capacity loss, keep at 0.7V for a prolonged period of time (24 hr).

Usual Applications

High capacity, long usage, low cost.

Replacement of nonrechargeable alkaline AA and AAA cells in consumer electronics and toys due to similar capacities and voltage range, while being rechargeable.

Hybrid electric cars.

Almost completely replaced by Li-ion in notebooks, mobile phones, and PDAs due to lower energy density.

Development Trends

AB$_2$-type alloys have higher capacity but are more expensive. Starting to appear in applications where high energy is required.

Various catalysts have improved low-temperature performance, power capability, and self-discharge performance.

Some AA cells (brand name Eneloop^{TM}) are now being sold precharged and ready to use due to significantly improved self-discharge behavior.

Lithium Ion.
Active Materials and Reactions.

1.5.4   Lithium Ion Battery

Discharge

Minimum 2.7V/cell (2.5V for LiFePO$_4$ and 1.5V for lithium titanate).

Li ions leave carbon (or other intercalation anode) matrix and are adsorbed by delithiated LiCoO$_2$ (or other intercalation cathode) matrix.

Charge

Typically 4.2V (lithium cobalt oxide) or 4.35V for improved electrolyte cells. Can be 3.6V (LiFePO4) or 2.7V (lithium titanate).

Li ions are extracted from LiCoO$_2$ (or other intercalation cathode) and are intercalated into carbon (or other intercalation anode) matrix.

Reactions

See Table 1.4 for lithium-ion battery reactions.

The overall intercalation reaction is illustrated in Figure 1.22.

1.5.4.1   Construction

The mechanical designs used for Li-ion batteries can be summarized as cylindrical, prismatic, and laminated (pouch) cells. All designs have common characteristics:

- Lithium-carbon intercalation compound is extremely active, therefore only organic electrolytes can be used.
- Contact of interiors with air or water destroys the battery, therefore casing and vent are designed to be extremely water and air resistant.
- All cells use a microporous separator that prevents growth of Li-dendrites and shuts down discharge in cases of overheating.

Table 1.4
Lithium-Ion Battery Material Options and Reactions

Positive Electrode (Several Options)	Negative Electrode (Several Options)	Organic Electrolyte	Discharge → / Charge ←	Positive Electrode	Negative Electrode	Electrolyte
Li$_{1-x}$CoO$_2$ Li$_{1-x}$Co$_{1/3}$Ni$_{1/3}$Mn$_{1/3}$O$_2$ Li$_{1-x}$Co$_{0.2}$Ni$_{0.8-x}$Al$_x$O$_2$ Li$_{1-x}$Mn$_2$O$_4$ Li$_{1-x}$FePO$_4$	Li$_x$[C$_6$] Li$_{4+x}$Ti$_5$O$_{12}$ Li$_x$Si	EC/DMC, +LiPF6	Discharge →	Li$_2$CoO$_2$ LiCo$_{1/3}$Ni$_{1/3}$Mn$_{1/3}$O$_2$ LiCo$_{0.2}$Ni$_{0.8-x}$Al$_x$O$_2$ LiMn$_2$O$_4$ LiFePO$_4$	[C$_6$] Li$_{4+x}$Ti$_5$O$_{12}$ Si	Same
Delithiated form of: lithium cobalt oxide (LCO) NMC NCA Manganese spinel Lithium iron phosphate	Lithium intercalated into carbon, lithium titanate, or forming an alloy with Si	Ethylene carbonate, methylene carbonate and many others. Lithium gexaflourophosphate as conductive salt	Charge ←	Lithium cobalt oxide (LCO) NMC NCA Manganese spinel Lithium iron phosphate	Graphite or hard carbon Lithium titanate Silicon	Same

Positive electrode

Negative electrode

Charge

Li+

Co
O

Li

Li+

Discharge

LiCoO₂

Specialty carbon

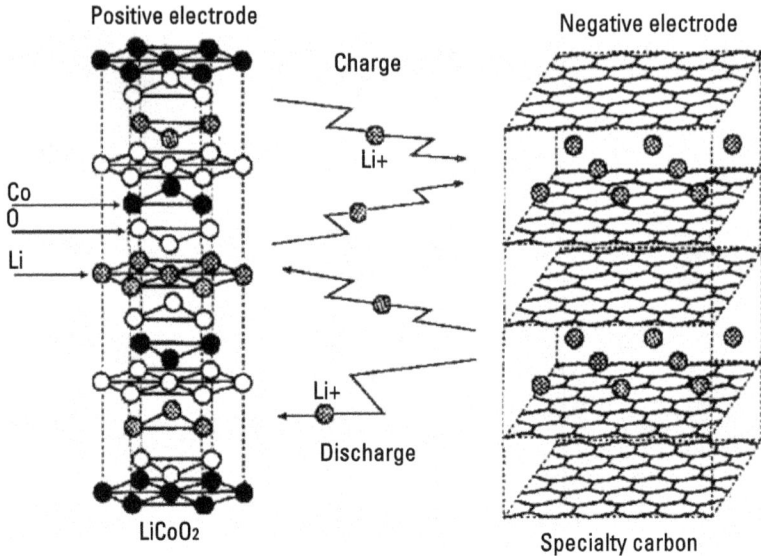

Figure 1.22   Crystalline structure and schematic charge/discharge reactions of Li-ion intercalation materials.

- Electrodes use thin foils as current collector. Anode is Cu foil, and cathode is Al foil.

- Active materials are slurry coated on the surface of current collectors as 50- to 100-μm thin films.

The cylindrical Li-ion design is similar to that of a NiCd battery. Electrodes are tightly wound together with a separator and placed in a robust steel casing. These batteries typically include a safety device (PTC) that shuts down discharge if the current or temperature exceeds a predefined threshold (typically 2C rate and 75°C). The most common format is the 18650 cell, which has an 18-mm diameter and 650-mm length. The capacity of an 18650 cell varies from 2.2 to 3 Ah depending on the grade.

The prismatic design is similar to the cylindrical design except that after winding the roll is compressed and placed in a prismatic aluminum casing. In some cases electrodes are stacked instead of winding. This provides for better power capability, but is more labor intensive.

The laminate (pouch) design is similar to the prismatic design except that wound or stacked electrodes are glued together using either a thermoplastic separator or gelating agents that are added to electrolyte. This gives the electrode stack mechanical stability even without external pressure, therefore no robust casing is needed. A flexible Al-coated polypropylene pouch is used instead of a casing. This type of cell used to be called a Li-polymer battery; however, because polymer electrolytes were found to have an inferior rate capability compared to liquid electrolytes, only the pouch design was preserved, and primarily liquid electrolytes are used.

1.5.4.2 Charging Method

Li-ion battery charging is performed by the CC/CV method, illustrated in Figure 1.23.

Charge Voltage

The voltage between the charging terminals should not be higher than specified charging voltage after taking into account fluctuation in power supply voltages and temperature deviations; it should be no more than 4.20V. [Set this at 4.20V (max) after taking into account fluctuations in power supply voltages, temperature deviations, etc.] The charge voltage varies according to model cell chemistry. Cells with coke-based anodes use 4.1V and there are many possibilities depending on cathode chemistry. Specifically $LiFePO_4$ cells will use 3.6V and lithium titanate cells 2.7V. Some new improved electrolytes can allow voltages up to 4.35V; they are likely to be found in the newer, higher energy cells. Check the manufacturer's specifications for charging. It is *extremely important not to exceed the charging voltage* because exceeding it will cause accelerated cell degradation or even an explosion.

Cell Voltage Control

In case of a state-of-charge (SOC) misbalance between serially connected cells, some cells can reach their maximal charging voltage earlier than others. Therefore, it is important to monitor each cell voltage separately, and terminate charging of the pack when the highest cell voltage reaches an unsafe region

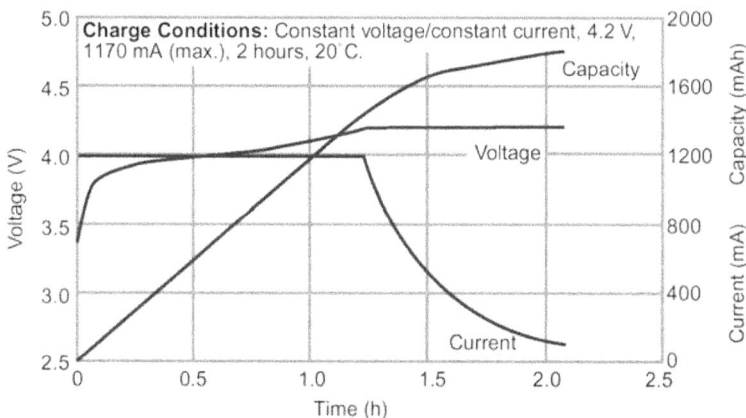

Figure 1.23   Charging profiles of a Li-ion battery cell using the CC/CV method.

(4.25 to 4.3V). Overcharging of even a single cell above this voltage will cause the entire pack to explode. Industry guidelines (1625-2008, *IEEE Standard for Rechargeable Batteries for Multi-Cell Mobile Computing Devices*) require that two independent devices monitor cell voltages and terminate charging if the highest voltage is exceeded by any cell. To ensure fault-redundant protection in a Li-ion battery, the pack design includes several independent overvoltage, overcurrent, and overtemperature protection circuits. An example of such a design is given in Figure 1.24.

Cell Balancing

To prevent cell voltages from becoming too high and causing premature termination, which results in a battery pack being undercharged, the SOC of the cells connected in series can be equalized by a cell-balancing algorithm implemented on the battery pack side. All gas gauging ICs of the bq series such as bq30z55 offer cell-balancing features without any additional external components.

Charge Current

Current should be limited to the 1C rate to prevent overheating and the resulting accelerated degradation. However, cells designed for high power capability can sustain higher charge rates. The rate should be selected so that at the end of the charging period, the temperature does not exceed 50°C.

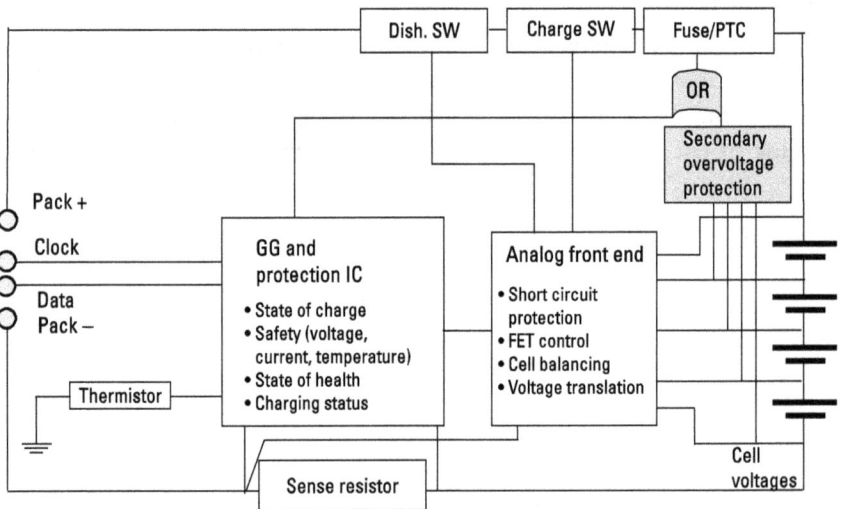

Figure 1.24   Block diagram of an example battery pack protection.

Temperature of the Battery Pack During Charge

Ambient temperature should be from 0° to 45°C. Lower temperatures promote formation of metallic Li, which causes cell degradation. Higher temperatures cause accelerated degradation because they promote a Li-electrolyte reaction.

Low-Voltage Battery Pack Charge

When the voltage per cell is 2.9V or less, charge using a charge current of 0.1C rate or less. This is to provide recovery of the passivating layer, which might be dissolved after prolonged storage in the discharged state.

A low rate charging also prevents overheating at 1C charge when previous overdischarge might have dissolved Cu from the anode current collector, which can cause shorted electrodes as a result of the Cu deposition. If the voltage during low-rate charging does not increase above 3V after about 30 min, then a short is present. The charger should detect this condition by observing that the charge is below 3V and indicate a fault if the maximal time has been exceeded. Most chargers as well as battery gauges include a precharge timer that will ensure safe termination in such case.

Termination of Charging

The system will determine when a battery is full by detecting the charge current. Stop charging once the current has reached the C/20 rate. This value can vary among cell manufacturers. Higher termination current means faster charge termination at a cost of lower charge energy. In some systems where charge current is reduced to currents below C/20 (e.g., due to extremely large battery size), charge termination can be done when the maximal charging voltage (typically 4.2V/cell) is reached.

Charge Timer

A total charge timer and a charge completion timer should be included to prevent unlimited charging and overheating in case of a microshort. These devices can be either part of a gauge or a charger. Usually charging time can be limited to 3 to 5 hr. Allowing charging to continue for a longer time will provide a longer battery pack life, because charging becomes slower with aging.

1.5.4.3   Rate Capability

The rate capability of Li-ion batteries can be tailored within a wide range by using thicker or thinner layers of active materials. Thinner layers result in a higher rate capability but less energy density. The typical 18650 cylindrical cells used in notebooks are designed for maximal C-rate discharge. However, cells rated for 10C discharge are used in portable power tools, and even cells capable to 60C rate discharge are used for power backup/regenerative breaking in hybrid gas-electric vehicles. Some cathode materials, such as Li-iron phosphate, manganese spinels, and anode materials, such as lithium titanate, allow for higher rate capabilities.

Rate capability is severely degraded at temperatures below 0°C because of the low conductivity ability of organic electrolytes. Some electrolytes are better than others; therefore, referring to manufacturers' data on low-temperature discharge is important.

1.5.4.4　Healthy Treatment

Healthy Habits

Most stable in 50% charged state.

High voltages accelerate corrosion and electrolyte decomposition. Charging should be limited to the maximal voltage specified by the manufacturer (most commonly 4.2 or 4.2V).

Short deep discharge is not detrimental, but extended storage in the discharge state results in dissolution of the protective layer and resulting capacity loss.

High temperature is main killer. Provide appropriate cooling and place battery far from heat-generating circuits. Take battery out of notebook if used as desktop to prevent exposure of the pack to high temperatures.

Use battery soon after manufacture. Discharge capacity degrades even if not used.

Storage at low temperatures increases shelf life.

If used for a standby application, charger should terminate charging and not resume until SOC drops below 95%. Enforcing constant voltage after full charge is reached is not recommended because the parasitic reaction rate is drastically higher close to maximal charge voltage, and letting battery voltage relax slightly below it makes a difference in degradation rate.

Unnecessary charging or discharging should be avoided; unlike with NiCd and NiMh batteries, no benefit is gained from "exercising" the battery.

Degradation Mechanisms

Reaction of Li-carbon compound with electrolyte. Despite the SEI protective layer, this reaction is always ongoing although very slowly and is accelerated by high voltage and high temperature. This reaction is making some of the lithium inaccessible in the form of insoluble compounds. These compounds also block the pores, electrically disconnect particles of the active material, and increase the resistivity of the SEI layer, increasing overall battery impedance.

Cycling the battery causes expansion/contraction of active materials, which causes cracks in the SEI layer, which then has to regrow. Therefore, deep cycling does accelerate degradation 5 to 10 times compared to storage during the same time period. Shallow cycles that do not cause significant expansion do slightly increase degradation, but mostly due to temperature rises.

Electrode corrosion. Very thin Al and Cu foils are used as current collectors. They are prone to corrosion, particularly at high states of charge. For this reason and electrolyte decomposition, batteries should be stored at 50% SOC.

Overdischarge at low rates below 2V OCV will cause dissolution of the Cu foil, and should be prevented by an electronic protector. However, if the voltage drops below 2V only momentarily due to a high current spike, most of the voltage decrease is due to an IR drop across the cell and the actual OCV remains in a safe range. For this reason, protectors typically have a delay of from tens of milliseconds to seconds before reacting to undervoltage. For the same reason, cells designed for high rate applications, such as $LiFePO_4$ cells, typically specify a lower minimal discharge voltage compared to cells designed for a low rate of discharge.

Usual Applications

High capacity, long usage.

Commonly used in digital cameras, video cameras, notebooks, tablets, mobile phones, plug-in hybrids and electric vehicles, power tools, power backup systems (UPS, telecom), and marine battery applications.

Development Trends

Since 1990 when Li-ion was first introduced, a capacity growth trend of 7% capacity/year has continued.

The initially used $LiCoO_2$ cathode is increasingly being replaced with Ni and Mn containing cathodes (NMC and NCA) due to their lower cost but similar energy benefits. These cells are mostly used in notebooks and other multiple-cell applications (automotive, power tools) because of their relatively lower voltage in the deeply discharged state, which makes their use less practical in single-cell electronics such as cell phones that have a 3.4V termination voltage in many cases.

A $LiFePO_4$ cathode provides much better safety than other types of cathodes—a three times higher thermal runaway temperature and a much lower reaction energy with electrolyte. Lower reactivity allows these materials to be prepared in nanoform, which significantly increases their rate capability, resulting in an approximately five times longer than usual cycle life. This chemistry has a very flat voltage profile due to two-phase intercalation, which is beneficial for applications that use unregulated power supplies, such as solar lights and power tools, ensuring even power delivery. Because of their lower voltage, such batteries have about half the energy density of the traditional $LiCoO_2$-based cells.

A lithium titanate anode provides even better rate capability and cycle life, but due to high cost and again lower energy density due to lower voltage, it is only suitable for niche applications where long life and high rate capability are absolutely critical. Can compete in many areas with supercapacitors.

Recent development of silicon anodes promises about a 30% increase of energy (for example, the capacity of 18650 cells would increase to 3.4 Ah). This is again at a cost of somewhat lower voltage in the deeply discharged state, so multiple-cell applications will be targeted first.

Cycle capability of Si-anode has been historically worse than that of carbon so it has to be seen if recent improvements will make these cells practical.

Next in line are high-voltage cathode materials that will allow energy density to be brought to a new level (30%+ improvements). They include:

Over-lithiated oxides (OLO), $xLi_2MnO_3 \cdot (1 - x)LiMeO_2$. Note that the term *over-lithiated* is somewhat misleading because it sounds like material is less oxidizing. In fact the opposite is true. Because Li is replacing some of the Mn while the same amount of oxygen remains, the overall oxidation state of an active metal such as Mn increases from +3 to +3.5, giving this material higher voltage and higher energy density.

Mixed cobalt/manganese/nickel/iron phosphates, such as $Li_xCo_{0.8}$ $(M1_aM2_bM3_c)_{0.2}PO_4$, operate at voltages up to 4.85V.

Development of new electrolytes and electrolyte additives will enable high-voltage operations. This is an active area of research that will enable future energy increases.

More futuristic developments that are not likely to be commercialized within a 5-year time frame include the following:

Lithium-sulfur battery: Despite much higher theoretical energy density, this chemistry does not allow for good cyclability due to solubility of Li-

Table 1.5

Battery Characteristics

Chemistry	Nominal Voltage (V)	Maximum Charging Voltage (V)	Mass Energy Density (Wh/kg)	Volume Energy Density (Wh/L)	Max Discharge Rate
Lead-acid	2	2.7	36	103	1–10C
NiCd	1.2	1.6	46	200	10–20C
NiMH	1.2	1.6	69	246	1–2C
Li-ion	3.6	4.2	135	365	2–10C

polysulfides, which results in loss of cathode integrity. Some new developments involving carbon nanofibers might allow a breakthrough, but cost might prove prohibitive.

Lithium-air battery: This highest theoretically possible energy density battery will first find application as a primary battery due to daunting challenges in achieving reversibility and good cycle life in presence of air impurities. Low rate capability due to the need to provide electron exchange with a gas-phase oxygen (require expensive catalysts and large porous electrodes) will initially limit it to niche applications where high energy is a must but power is not critical (for example, hearing aid batteries).

1.5.5 Battery Chemistries Overview

Table 1.5 provides an overview of some of the characteristics of the battery types discussed above.

References

[1]   Barsukov, E., and J. Ross Macdonald (Eds.), *Impedance Spectroscopy: Theory, Experiment, and Applications, 2nd ed.,* New York: John Wiley & Sons, 2005.

[2]   Tobishima, S., J. Yamaki, and T. Hirai, *J. of Applied Electrochemistry,* Vol. 30, 2000, pp. 405–410.

2

Battery Charger Techniques

There are many kinds of rechargeable battery chemistries. Among them, lead-acid, nickel-cadmium (NiCd), nickel metal hydride (NiMH), lithium-ion (cobalt based), and lithium-iron-phosphate (LiFePO$_4$) rechargeable batteries are the most popular for many different applications. However, their battery charging algorithms are different in order to optimize battery charging performance. In this chapter, battery charging algorithms, battery charging system topologies, design considerations, the main battery charger applications, and popular battery charger ICs are discussed.

2.1   Lead-Acid Battery Charger

A lead-acid battery charger usually has two main tasks to achieve. The first is to charge the battery such that capacity can be restored as quickly as possible. The second is to maintain the battery's capacity by compensating for capacity loss due to self-discharge by applying a constant voltage to a fully charged battery. Because battery voltage is a function of its cell temperature with –3.9 mV/deg per cell, the charge voltage needs to be adjusted according to its cell temperature. If such temperature compensation is not considered, loss of capacity could happen below the nominal design temperature, and overcharging with a subsequent degradation of cycle life could occur at higher temperature.

Figure 2.1 shows the typical lead-acid battery charge profile, which comprises four charge phases:

Phase 1: Trickle charge or preconditioning. If the battery cell voltage is below 1.7V per cell at 25°C, it is considered to be a deeply discharged battery. A small trickle charge current of $I_{TRICKLE}$ is applied to bring the

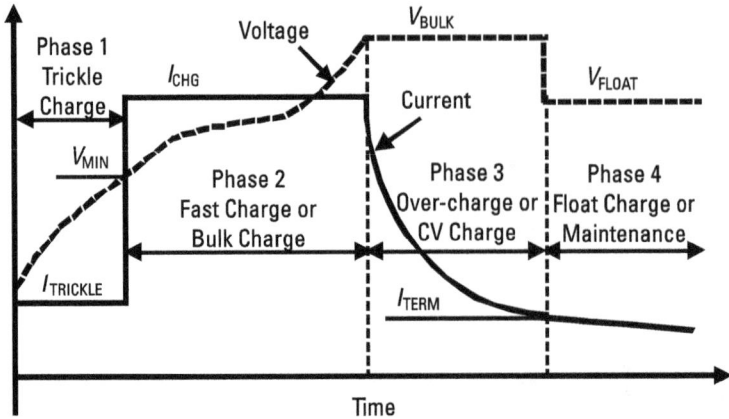

Figure 2.1 PbSO4 battery-charging profile.

battery voltage up to the predetermined threshold of V_{MIN} (e.g., 1.7V per cell) to determine if the battery cell is internally shorted or not. Only after the cell reaches such a voltage threshold is a fast charge current applied to the battery.

Phase 2: Fast charge or bulk charge. The charger moves into the fast-charge phase with a high charge current rate of I_{CHG} when the battery voltage exceeds the trickle charge voltage threshold. The battery voltage increases for restoring battery capacity. About 70% capacity can be charged during this charge phase.

Phase 3: Overcharge or constant voltage charge. When the battery voltage reaches the bulk-charge voltage threshold of V_{BULK}, a constant bulk voltage is applied to the battery. The charge current then gradually decreases to the termination current of I_{TERM}, typically about one-tenth of the fast-charge current or bulk-charge rate, where the overcharged is terminated.

Phase 4: Float charge or maintenance charge. To maintain full capacity, a fixed float voltage of V_{FLOAT} is applied to the battery. This float charge is used to compensate for the capacity loss due to self-discharging. When the battery voltage drops to 90% of the float voltage, the charger resumes its fast-charge stage. The float voltage is typically around 2.2 to 2.3V per cell depending on its temperature.

At higher temperatures, the fast-charge current for lead-acid batteries should be reduced according to the typical temperature coefficient of 0.3% per degree centigrade. The maximum temperature recommended for fast charging is about 50°C, but maintenance charging can generally proceed above that temperature. Some lead-acid battery ICs can minimize the solution size. Such ICs include the bq2031and bq24450 lead-acid chargers from Texas Instruments. The MAX8724 and bq24725A host-controlled SMBus switch-mode charger is used to achieve any charge algorithm with a microprocessor.

2.2 NiCd and NiMH Battery Charger

2.2.1 Nickel-Based Battery Charge Characteristics and Charge Profile

NiCd and NiMH batteries have their advantages in some power applications, although the Li-ion battery is more popular because of its high energy density and high cell voltage. Both of these characteristics are advantageous in portable equipment. Nickel-based rechargeable batteries are capable of a high discharge rate because they have lower internal impedance, good cycle life, do not require special safety protection requirements, and are less expensive than Li-ion batteries. Between these two chemistries, NiMH batteries have greatly replaced NiCd batteries because they have a 40% to 100% higher capacity and NiCd batteries raise concerns in terms of their toxic cadmium content.

Charging the nickel-based batteries is quite simple from a circuit standpoint. A constant current is simply applied to them with a certain charge termination point. How much charge current is required to charge the batteries depends on how quickly the charge is needed. Of course, safe as well as fast charging is attractive for many applications, but we have to refrain from overcharging, which degrades the battery faster. Fast charging requires high current. One of the challenges is the heat generated by the internal resistance of the batteries when a high charging current is used. For a deeply discharged nickel-based battery, the charge efficiency that results from converting electrical energy into electrochemical energy is pretty high and the heat generated during this charge period is small. The cell voltage increases when a constant charge current is applied to a discharged battery. However, when it gets close to its full charge and the cell cannot hold more energy, the charging process becomes inefficient as some of the electrical energy cannot be converted to electrochemical energy and translated into heat, which could result in excessive temperature rise and gas venting to the outside atmosphere, especially at the end of charging. However, using the absolute cell temperature as the charge termination temperature is undesirable because cell temperature is not only related to the heat generated by the charging process, but also to the ambient temperature. The ambient temperature change rate is very slow, and the temperature rise rate could be quite a bit higher due to the excessive heating at the end of charging. Therefore, the cell temperature rise rate becomes an important charge termination parameters that should be monitored. In addition, the cell voltage equals the summation of the battery cell open circuit voltage and voltage drop across the battery internal resistance, as given by

$$V_{CELL} = V_{OC} + I_{CHG} \times R_{CELL}$$

where V_{OC}, I_{CHG}, and R_{CELL} are the battery cell open circuit voltage, charge current, and battery cell internal resistance, respectively.

The battery internal resistance is related to the battery temperature. The higher the battery cell temperature, the lower the battery resistance. After the cell voltage reaches its peak value, the cell voltage does not increase any more at the end of charging; instead it decreases as the voltage drop across the battery internal resistance becomes smaller. So we cannot terminate charging purely based on the cell voltage. Because most of the energy is converted to heat and the cell temperature increases very fast at the end of the charging period, typical charge termination is based on a temperature rise rate ($\Delta T/\Delta t$) of more than 1°C per minute, peak cell voltage, and cell voltage drop. Such a termination algorithm requires a high enough charge current, which typically needs to be higher than a 0.5C charge rate or the voltage drop cannot be big enough to be detected. Figure 2.2 shows nickel-based battery charge characteristics. It indicates that the rapid temperature rise occurs first, and the cell voltage reaches its peak voltage and then drops. The temperature rise causes the battery cell voltage drop.

Another charge method is to use a small charge current such as a 0.1C charge rate and below with a safety timer. Such a small charge current cannot generate excessive heat and ideally it takes 10 hours to fully charge the battery with 100% charge efficiency. In reality, it takes about 14 hr to fully charge

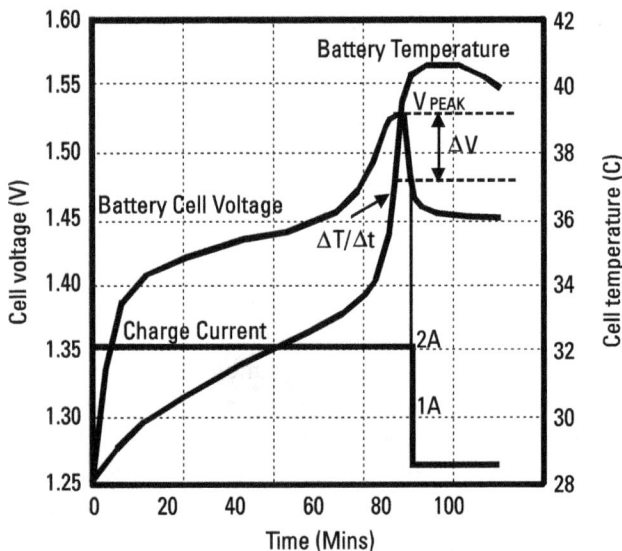

Figure 2.2   The nickel-based battery charge characteristics with 2A charge current.

the battery because of the charge efficiency. To ensure a safe battery charging system, a safety timer is usually used. When the charge safety timer expires, the charger automatically stops charging and provides a warning signal to the system that the battery may not be good enough.

2.2.2 NiMH Battery Charger Design Example

Figure 2.3 shows a 2A fast switch-mode synchronous switching charger operating at 550 kHz. The charger controller is the LTC4011 from Linear Technologies. It integrates all of the functions needed to charge Ni-based batteries. This includes constant current control circuitry, charge termination, automatic trickle and top-off charge, automatic recharge, programmable safety timer, and multiple status outputs. Power path control allows the charger to charge the deeply discharged battery while simultaneously powering the system. Such a high level of integration lowers the component count, enabling a complete charger to occupy less board area. For a NiMH battery pack, it is preferable not to charge NiMH cells in parallel because doing so would mask the effects that are used to determine end of charge. Initial battery charge qualification verifies that the battery voltage and battery temperature are within an acceptable range before charging at full current. For a deeply discharged battery, a low-current

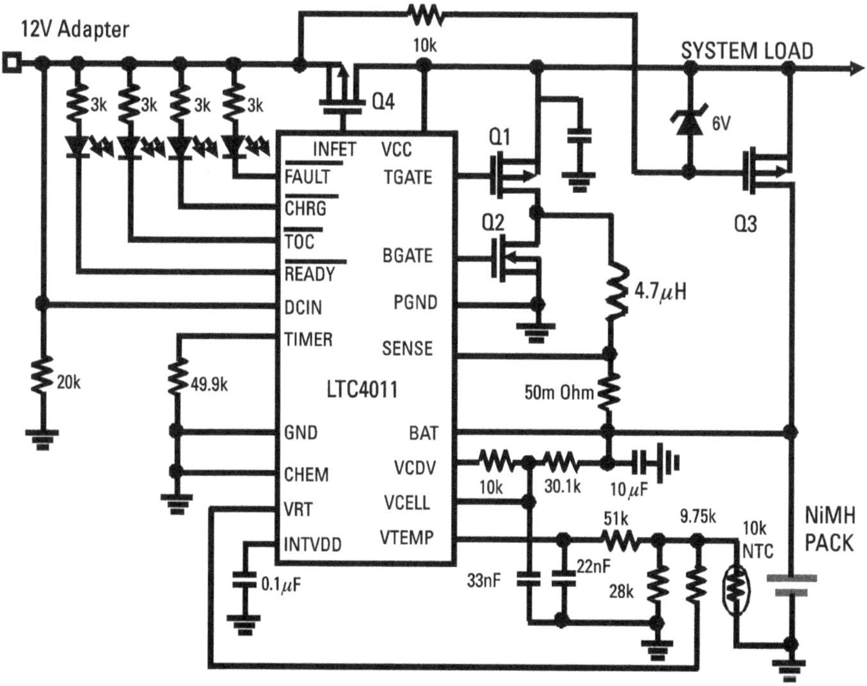

Figure 2.3 2A 1–16 cell NiMH battery charger application circuit.

trickle charge is applied to raise the battery voltage to an appropriate level before applying full charge current. When qualification is complete, the full programmed constant current begins. The charge termination methods used by this design example utilize battery voltage and battery temperature changes to reliably indicate when a full charge has been reached as a function of the charge current selected. The charge current has to be sufficiently high (between a 0.5C and 2C charge rate) to exhibit the voltage and temperature profile required for proper charge termination. After the charge cycle has ended, the charger continues to monitor the battery voltage. If the voltage drops below a fixed rechargeable threshold either due to an external load on the battery or self-discharge, the charger begins to recharge the battery with the charge termination algorithms immediately enabled.

A few NiCd and NiMH battery charger controllers such as the bq2002 from Texas Instruments and MAX712 from Maxim Integrated Circuits are available for effectively charging NiCd and NiMH batteries.

2.3   Li-Ion and Li-Polymer Battery Charger

2.3.1   Li-Ion and Li-Polymer Charge Characteristics and Principle

Figure 2.4 shows a widely used constant current (CC) and constant voltage (CV) charge profile for a Li-ion battery. Most dedicated Li-ion battery charge ICs are designed to charge the battery in this manner. The charging of a Li-ion battery consists of three phases: precharge, fast-charge CC, and CV. In the precharge phase, the battery is charged at a low rate (typically about one-tenth of the fast-charge rate) when the battery cell voltage is below 3.0V ($LiCoO_2$

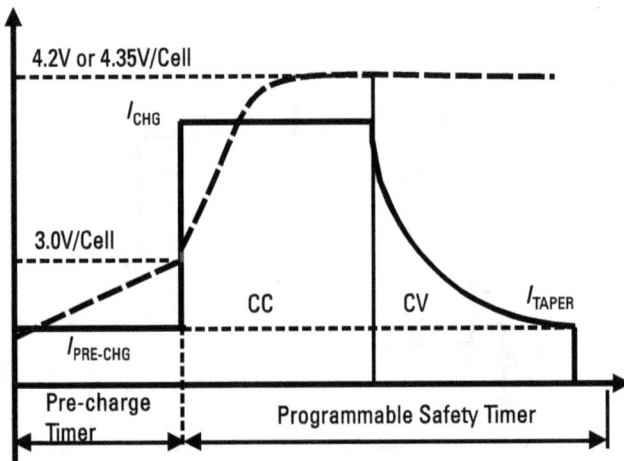

Figure 2.4   Li-ion battery CC-CV charge profile.

cathode material). This provides recovery of the passivating layer, which might have dissolved after prolonged storage in a deep discharge state. It also prevents overheating at a high charge rate when partial copper decomposition appears on anode-shorted cells on overdischarge. This precharge mode is also used to wake up a deeply discharged battery within a certain time period. A precharge safety timer prevents charging of a dead battery for a long time period. If the battery voltage cannot be charged to reach 3.0V per cell within this safety time period, then the battery is considered to be dead. Typically a warning signal is provided to tell the end user that the battery is no longer usable. When the battery cell voltage reaches the typical 3.0V, the charger enters the CC phase. Fast-charge current is usually limited to a 0.5C to 1C rate to prevent overheating and the resulting accelerated degradation. The charge rate should be selected such that the battery temperature does not exceed 45°C. The fast-charge current allows the battery to be charged very quickly and translates the electrical energy into electrochemical energy in the battery. The battery is charged at the fast-charge rate until the battery reaches a voltage regulation limit (typically 4.2V/cell for a $LiCoO_2$-based cathode; 4.35V for the combined Li-Ni-Mn-Co chemical compound and $LiCoO_2$-based cathode battery).

The charger starts to regulate the battery voltage and enters the CV phase while the charge current exponentially drops to a predefined termination current, at which point the battery charging is terminated. This charge current drop occurs naturally; that is, it is not controlled by the charger given the fact that a constant voltage provided by the charger is applied to a battery pack equivalent circuit of a battery internal resistance and a huge capacitor. The charge current is gradually reduced with the increase in the battery's open circuit voltage. The termination charge current is usually about 5% to 10% of the fast-charge current. A fast-charge safety timer is usually required to prevent a battery from being charged for an excessively long period. The battery charger must be terminated when the safety time expires even if the battery does not reach the termination current.

The battery charging time is a very important design parameter for end users. For a given 1C charge rate, a battery charger usually takes about 30% of the charging time, whereas it can charge 70% of its total capacity during the CC phase. In contrast, it takes about 70% of the total charging time while only charging 30% of its total capacity during the CV phase. This is because a battery is not an ideal power source, but it has an internal resistance. The lower the battery internal resistance, the shorter the battery charging time. In real applications, the battery pack resistance includes the on-resistance of two back-to-back protection MOSFETs in the battery pack, trace resistance and battery internal resistance. Such total resistance can be as high as a few hundreds of milliohms. Increasing the battery charge current rate can reduce the battery charge time in the CC mode while increasing the charge time in CV mode. At higher charge current, the transition between the CC and CV phases occurs earlier, because the cell voltage is higher due to the voltage drop across the cell's internal resistance, and therefore the CV voltage is reached sooner. Overall, however, increasing the charge current can reduce the battery charge time. However, a battery charge rate higher than 1C is usually not recommended because it impacts the battery's cycle life. Figure 2.5 shows the relationship between the battery cycle life and the battery charge rates for a Li-ion battery with $LiCoO_2$ cathode [1]. Usually the higher the battery voltage, the higher the battery capacity. However, a higher battery charge voltage results in a shorter battery cycle life. The battery cathode materials will react faster with the electrolyte at higher battery voltage than that of lower battery voltage. The cobalt material will be permanently lost during the chemical reaction. Thus, fewer energy storage materials become available, resulting in a battery chemical capacity loss. The battery cycle life reduces by half when it is charged to 4.3V although it can achieve about 10% more capacity initially. In contrast, when a battery is undercharged, its battery capacity will be lower. If the battery is undercharged (40 mV lower), it will lose about 8% of its total capacity. Therefore, battery charge voltage accuracy is very important.

Figure 2.6 shows the relationship between the charge current rate and cycle life [1]. The higher the battery charge current rates, the lower the battery cycle life. At high charge rate, the extra Li-ion will be deposited at the anode and it becomes metallic lithium when a free electron is available. The metallic lithium is a very active metal and easily reacts with the electrolyte, which results

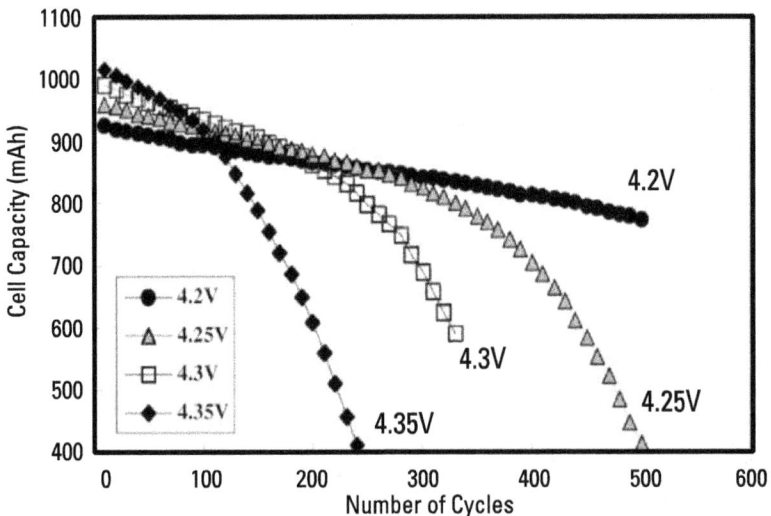

Figure 2.5 Relationship between battery charge voltage and cycle life for a Li-ion battery with a $LiCoO_2$ cathode.

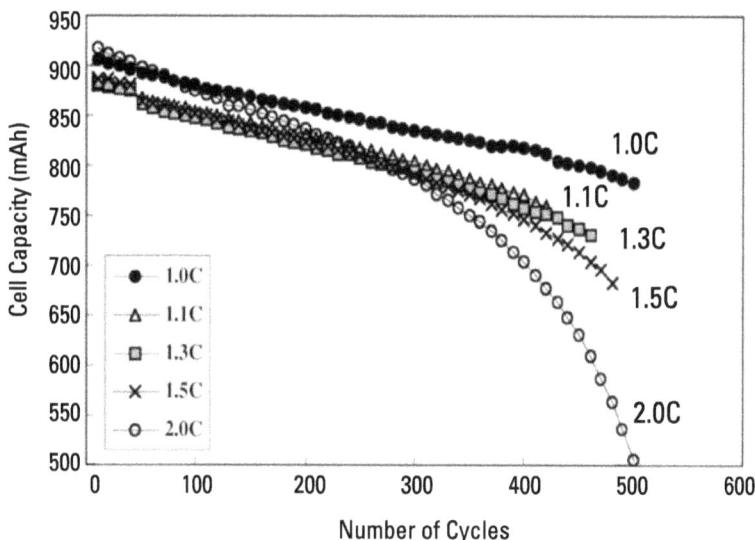

Figure 2.6 Relationship between battery charge current and cycle life for a Li-ion battery with a LiCoO$_2$ cathode.

in permanent lithium loss. As a result, the energy storage element, Li-ion, disappears, increasing the battery aging and, hence, reducing the battery cycle life. Therefore, the battery charge rate is recommended to be between 0.5C and 1C; typically a 0.7C charge rate is used for a LiCoO$_2$ cathode-based Li-ion battery.

2.3.2 Charge Temperature Qualification and JEITA Guideline

To safely charge a Li-ion battery, the battery can be charged only in a certain cell temperature range, typically between 0° and 45°C. Charging the battery at lower temperatures promotes formation of metallic lithium, which causes cell degradation. In contrast, charging the battery at higher temperatures is extremely dangerous and causes accelerated cell degradation as well. To further improve battery safety, the Japan Electronics and Information Technology Industries Association (JEITA) guideline was proposed. Figures 2.7 and 2.8 show the battery charge voltage and current limits for notebooks [2] and single-cell portable devices, respectively. Note that both charge voltage and charge current are functions of the battery cell temperature. The higher the cell temperature, the lower the charge voltage. When the cell temperature is above 45°C, the charge voltage is reduced in order to improve safety. This voltage in the figures is the maximum charge voltage including all tolerances. Note, however, that different cell manufacturers may have different specifications and we need to

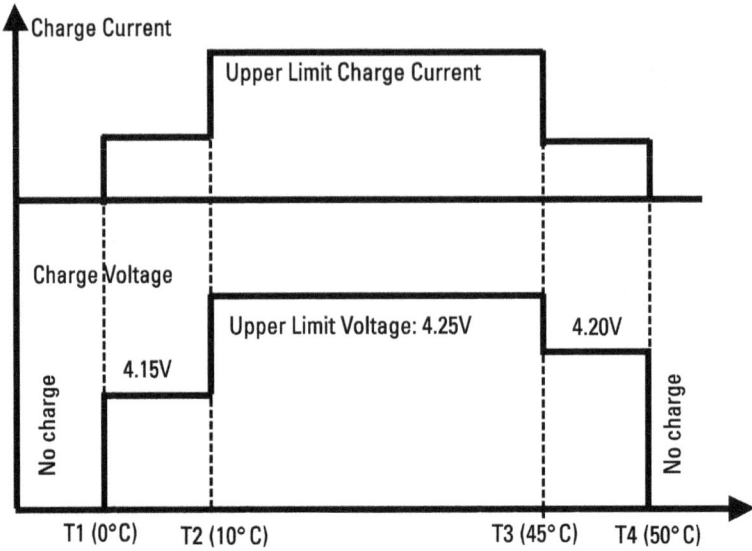

Figure 2.7 Battery charge JEITA guideline for notebook applications.

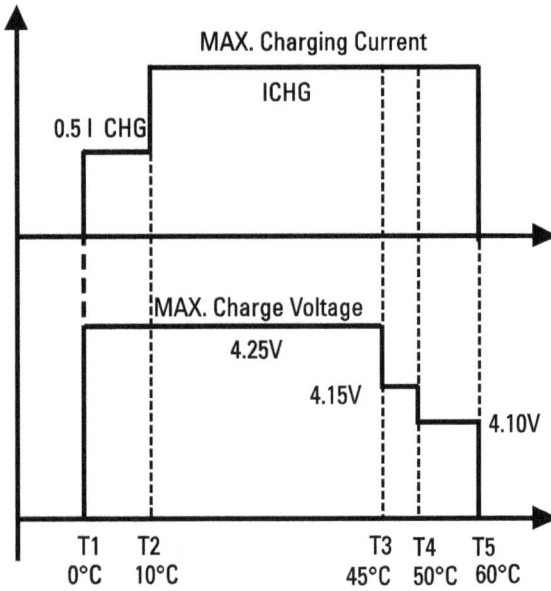

Single Cell Battery Charger JEITA Guideline

Figure 2.8 Battery charge JEITA guideline for single-cell applications.

follow the recommendations of the cell manufacturers' because cell chemistries differ from vendor to vendor.

2.3.3 Linear Battery Charger

The battery charger needs to accurately regulate both the battery charge current and charge voltage. The simplest charger is a low drop-out (LDO)-based linear battery charger. The major difference between a regular LDO and linear charger is that the linear charger has an additional charge current regulation loop for achieving constant output current besides the output voltage regulation. Figure 2.9 shows a block diagram for a linear battery charger. There are two regulation control loops: the charge current regulation loop and charge voltage regulation loop. The constant charge current and voltage can be achieved by adjusting the on-resistance of the pass element, MOSFET Q1, at different input voltages or battery voltages. The voltage drop across MOSFET Q1 always equals the voltage difference between the input voltage and battery voltage. The power dissipation across the MOSFET Q1 equals input voltage minus the battery voltage times the charge current, which is given by

$$P_{\text{loss}} = \left(V_{\text{IN}} - V_{\text{BAT}}\right) \cdot I_{\text{CHG}}$$

The battery charger power conversion efficiency is equal to the ratio of the battery voltage to the input voltage, and is given by

$$\eta = \frac{V_{\text{BAT}}}{V_{\text{IN}}}$$

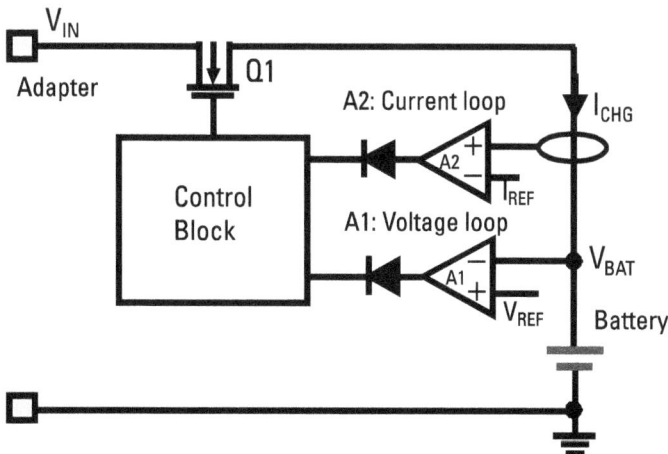

Figure 2.9 Linear battery charger block diagram.

The average efficiency is about 74% given the average battery voltage of 3.7V and input voltage of 5V and the fact that 26% of the input power is consumed and wasted by the pass element MOSFET Q1. Therefore, a linear charger is not suitable for high charge current or high voltage differences between input and output applications.

The main advantage is that it has a minimum number of components and is the smallest size and the lowest cost solution. Its drawback is the power dissipation across the power MOSFET Q1

If a 5V adapter is used to charge a 1,200-mAh or 2,200-mAh single-cell Li-ion battery, Figure 2.10 shows its power dissipation with a 0.7C rate fast-charge current. It has a maximum power dissipation of 1.68 and 3.0W when the battery transitions from the precharge to the fast-charge phase at a battery voltage of 3V, respectively. Power dissipation of 3.0W causes a 141°C temperature rise for a 3-mm × 3-mm QFN package with thermal impedance of 47°C/W. This exceeds the maximum silicon junction operating temperature of 125°C at an ambient temperature of 25°C. However, the thermal performance may not so bad because the battery voltage increases very rapidly at the beginning of the fast-charge mode because the battery has a relatively high internal resistance.

What are suitable applications for a linear battery charger? For a battery with a less than 800-mAh battery pack and 0.7C charge rate, the average power dissipation is about 700 mW, which could not cause significant thermal issues with a 5V input voltage and 3.7V average battery voltage. How do we achieve good thermal management for a linear battery charger? In a real linear charger design, the power switch MOSFET is typically integrated into the battery

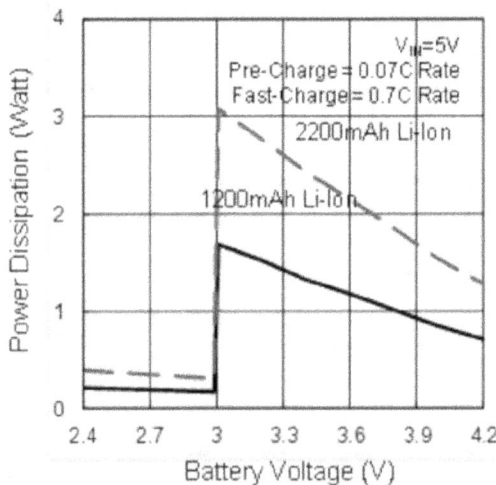

Figure 2.10   Power dissipation in a linear charger.

charger for minimize the solution size. But this increases the power dissipation into the charger. A thermal control regulation loop is usually introduced to regulate the maximum IC temperature around 125°C. The battery charger automatically reduces the charge current shown in Figure 2.11 when the battery charger IC junction temperature reaches the thermal regulation temperature and maintains the battery charger temperature in a safe region.

We now turn to a linear battery charger design example. A linear battery charger is very popular in low-end phones, headsets, and many other applications, where the battery capacity is less than 800 mAh and the battery charge current is less than 1A. Here are the battery charger design specifications:

Input voltage: 5V ± 10%

Battery capacity: 750 mAh

Fast-charge current: 750 mA

Precharge current: 75 mA

Fast-charge safety timer: 5 hr

Charge temperature qualification window: 0° to 45°C.

The bq24060 charger from Texas Instruments is designed to meet the above design specification and an example is shown in Figure 2.12. The resistor connected at TMR is used to program the fast-charge safety timer and the precharge safety timer is fixed at 30 min. ISET is used to program the fast-charge current while the charge voltage is fixed at 4.2V. The TS pin is connected to the thermistor, which monitors the cell temperature through a resistor divider for adjusting the battery charge temperature qualification window. STAT1 and STAT2 are used to indicate the battery charge status.

There are many stand-alone linear battery chargers with integrated power MOSFETs including the ISL6293 from Intersil, MAX8600A from Maxim

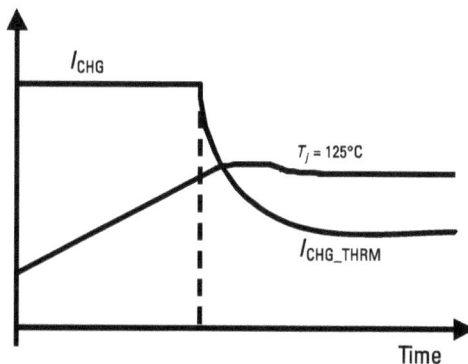

Figure 2.11   Thermal regulation by adjusting the charge current.

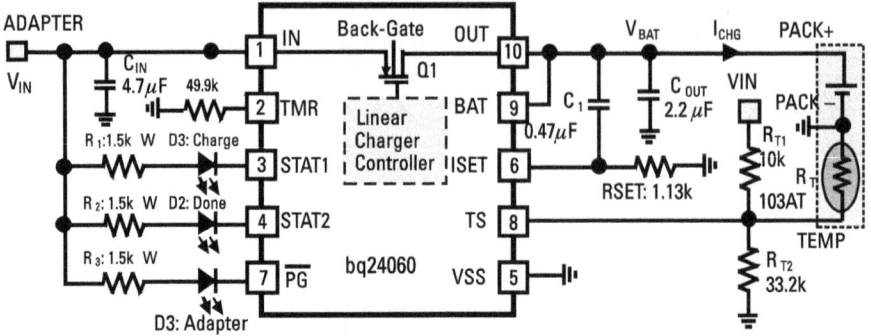

Figure 2.12 Li-ion linear charger design example.

Integrated Circuits, and LTC4057 from Linear Technologies. All of these integrated chargers could accurately regulate the charge voltage and current with a safety timer, charge indictors, and thermal regulation.

2.3.4 Switch-Mode Battery Charger

The main disadvantage of the linear battery charger is its lower battery charge efficiency, which has about 72% efficiency for charging a single Li-ion cell battery from a 5V power source. The charger itself dissipates 28% of input power.

How can you efficiently use available power to charge a Li-ion battery faster and minimize power dissipation for improving the thermal design? The synchronous switching converter has a higher power conversion efficiency than a linear regulator because the switcher operates in either a completely on or off mode, not in a linear operation.

Figure 2.13 shows a synchronous switching buck converter, and its switching waveforms are shown in Figure 2.14. Switchers Q1 and Q2 turn on and off complementarily to generate a switching waveform with amplitude of V_{IN} and duty cycle of D at the switching node. The inductor L and output capacitor C_o are composed of a low-pass filter to get rid of the high-frequency ripple for creating a DC output voltage. Because the average inductor voltage equals zero over one switching cycle in a steady state, the output voltage V_O is equal to the duty cycle of switcher Q1 times the input voltage, and is given by

$$V_O = DV_{IN}$$

for the duty cycle D = t_{on}/T_s, where ton is the turn-on time of switcher Q1 and T_s is the switching period, respectively.

How can we use the traditional switch-mode DC-DC buck converter as a battery charger? The battery charger needs a constant output current for

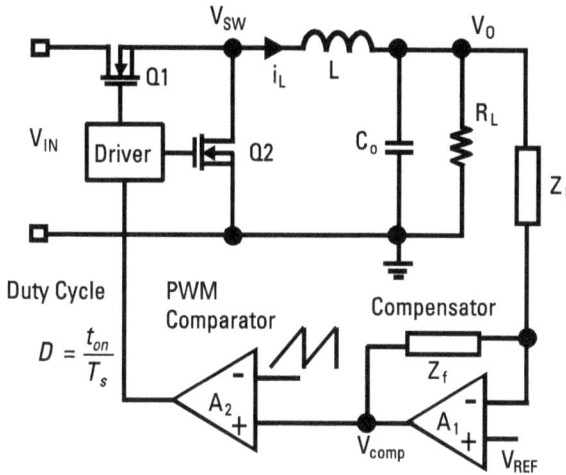

Figure 2.13   Synchronous switching DC-DC converter with voltage regulation feedback.

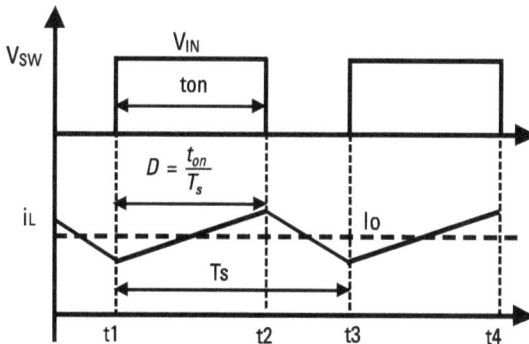

Figure 2.14   Switching waveforms of Figure 2.13.

charging the battery in addition to constant output voltage. We can implement the battery charger by adding an output charge current regulation loop in the traditional switch-mode DC-DC converter. Figure 2.15 shows a high-efficiency synchronous switching charger. Actually any existing switching mode DC-DC converters can be used as a charger by adding an additional output current regulation loop with additional safety functions such as a safety timer and charge termination.

For a limited input current power source from a USB port or adapter, the synchronous switching buck-based charger can provide a higher output charge current than a linear charger. Based on the power balance between the input and the output, the effective battery charge current I_{CHG} is given by:

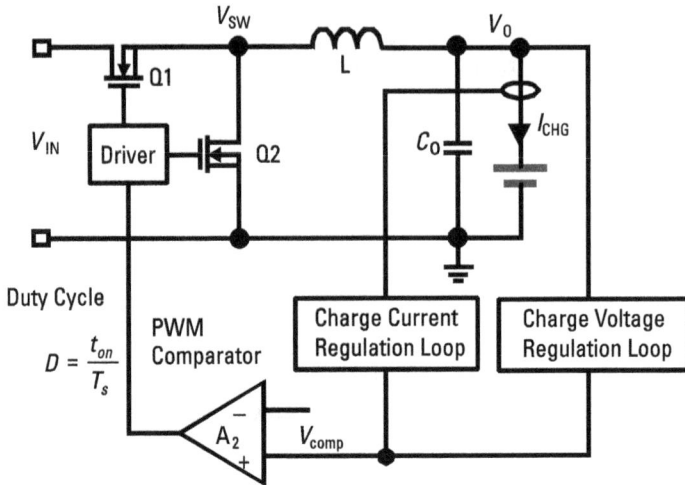

Figure 2.15   Synchronous switching buck-based charger.

$$I_{CHG} = \frac{V_{IN}}{V_{BAT}} \times I_{IN} \times \eta$$

V_{IN}, V_{BAT}, I_{IN}, and η are the input voltage, battery voltage, input current, and power conversion efficiency, respectively. This equation shows that it provides higher charge current at lower battery voltage, and lower charging current with a higher battery voltage. The charger current could be higher than the input current limit as the input voltage is always higher than the battery voltage. For USB charging applications, the input voltage is constant at 5V, and the input current is limited to either 100 or 500 mA. Thus, the charge current is inversely proportional to the battery voltage and power conversion efficiency. A practical step-down synchronous switching buck converter can achieve about 90% efficiency.

Figure 2.16 shows the battery charge current for a switch-mode charger and a linear-mode charger with 5V USB input and a 500-mA input current limit. It shows that the switching charger can provide 40% higher average charging current than a linear charger. Therefore, the switching charger can charge the battery faster than that of a linear charger. This is simply because the switching charger has less power consumption and more power available to charge the battery if the input power is fixed. For space-limited applications, the smallest charger is required, which may require the charger to have integrated power MOSFETs. Besides the charge voltage and charge current regulation and input current regulation loops, the charger also needs a thermal regulation loop for the charger with integrated power MOSFETs in portable devices because

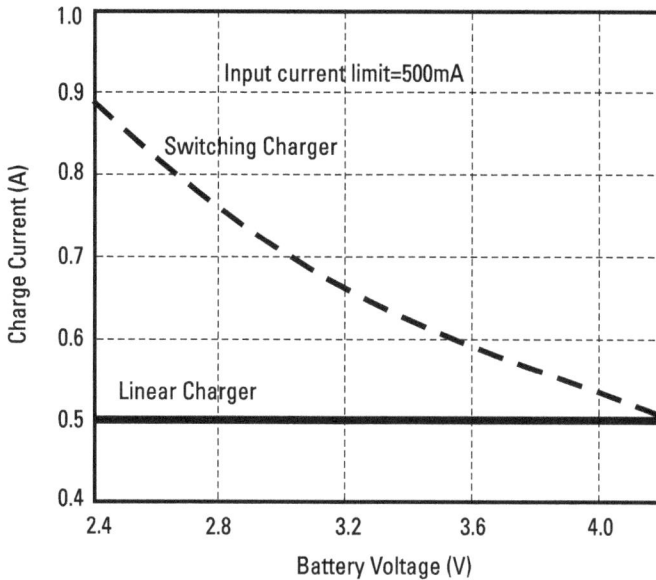

Figure 2.16   Battery charge current for a linear charger and switch-mode charger.

the power MOSFETs usually have high power dissipation, which results in much heat being generated and may exceed the silicon junction temperature.

2.3.5   Switch-Mode Battery Charger Design Example

Figure 2.17 shows one example of a 3-MHz synchronous switching step-down single-cell charger with integrated power MOSFETs to minimize the size. Operating at a 3-MHz switching frequency significantly reduces the passive component size of the output inductor and capacitor so that a chip inductor can be used in size-limited applications in portable devices. Q1 is an N-MOSFET that blocks any leakage current from the battery to the input when the adapter is not connected. It is also used to sense the input current to make sure that the input current will not exceed the current limit of the input power source. To drive the N-channel MOSFETs of Q1 and Q2, a charge pump circuit comprised of a bootstrap capacitor C4 and integrated diode is used to charge capacitor C4 when MOSFET Q3 conducts. The integrated type III loop compensator for battery charge voltage and current loop is used to further minimize the number of external components and improve the reliability.

For USB applications, the maximum input current is 500 mA for USB 2.0 and 900 mA for USB 3.0. When the input current reaches the maximum current limit set by the I²C, the input current regulation loop becomes active and reduces the PWM controller's duty cycle so that it meets with the USB

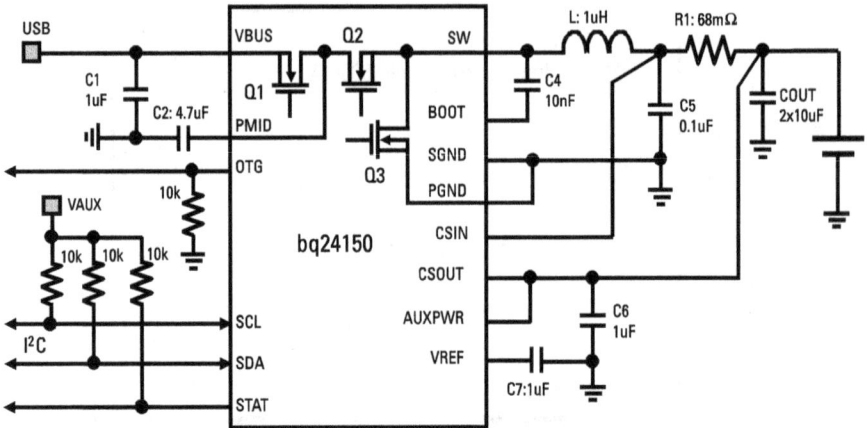

Figure 2.17   A 3-MHz synchronous switching single cell Li-ion battery charger.

specifications and avoids the USB crash. Input current limit regulation accuracy also plays an important role in order to maximize the power available from the USB port. The higher the input current regulation accuracy, the more input power can be drawn for charging the battery. This is mainly because the input current limit set point could be close to the maximum input current of the USB port with higher regulation accuracy. For example, with 10% input current regulation tolerance (90% accuracy), the input current regulation point can be set at 450 mA for a 500-mA USB port so that it never exceeds the 500-mA limit. However, with 5% input current regulation tolerance (95% accuracy), we can then set a 475-mA input current limit, which never exceeds the 500-mA USB current limit. Obviously, higher input current regulation accuracy is able to maximize the input power drawn from the USB port and charge the battery faster.

Figure 2.18 shows the experimental results for charging a Li-ion battery through a linear charger and a switch-mode charger through a USB port. It is shown that the switch-mode charger can shorten the battery charging time by more than 10%.

In the preceding discussion, the switch-mode buck converter is used as a charger, which requires that the input voltage be higher than the battery voltage. However, if the input voltage of the power source is lower than the battery voltage, then we have to use either boost or buck-boost-like Sepic converters for charging the battery. For example, a 5V input power source is used to charge two series Li-ion cell battery packs. Buck-boost charger LT1512 from Linear Technologies employs a Sepic converter for achieving the battery charging function.

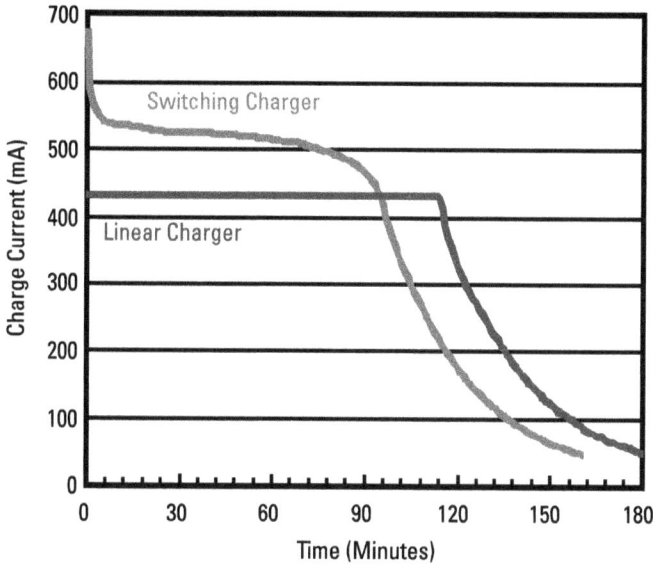

Figure 2.18 The battery charging time of a linear charger and switch-mode charger.

2.3.6 USB Battery Charging

USB ports have become very popular for providing power to portable devices. Recently the power from not only the adapter but also from the USB port has been used to charge the battery. It is extremely convenient and handy for consumers to just carry a USB cable for charging portable devices while traveling.

The most useful benefit of a USB's power capabilities is the ability to charge batteries in portable devices. A practical USB battery charger optimizes battery charging performance, safety, and the user experience. The USB specification spans several generations of power management. The initial USB 1.0 and 2.0 specifications described two types of power sources: 5V at 500 mA and 5V at 100 mA for powering connected devices. These specifications were not written with battery charging in mind; they were intended only to power small peripherals like mice and keyboards. This did not prevent designers from finding a USB battery charging solution. The recent development of a supplementary USB specification [*Battery Charging Specification*, Revision 1.1, April 15, 2009 (BC1.1)], acknowledges that charging ability and describes power sources that can supply up to 1.5A. All USB power ports, when active, were classified as either "low power" (100 mA) or "high power" (500 mA). Any port could also operate in a suspended mode, which limits supply to 2.5 mA. Ports on PCs and laptops are "high power," whereas ports on hubs that receive no power other than what is supplied by the upstream USB host are considered "low power." Once plugged in, a device is initially allowed to draw up to 100 mA while enumerating and negotiating its current budget with the host. Subsequently it could raise 500 mA, or it might be held at 100 mA.

Standard downstream port (SDP): This is the same port defined by the USB 2.0 specification and is the typical form found in desktop and laptop computers. The maximum load current is 2.5 mA when suspended, 100 mA when connected and not suspended, and 500 mA when configured for that current. A device can recognize a SDP with hardware by detecting that the USB data lines, D+ and D–, are separately grounded through 15 kΩ. It needs to enumerate to be USB compliant, although much of the present-day hardware does not enumerate.

Charging downstream port (CDP): This defines the higher current USB port typically used with PCs, laptops, and other hardware. The CDP can supply up to 1.5A. A device plugged into a CDP can recognize it as such by means of a hardware handshake implemented by manipulating and monitoring the D+ and D– lines.

Dedicated charging port (DCP): This describes power sources like wall adapters that do not enumerate so that charging can occur without communication at all. DCPs can supply up to 1.5A and are identified by a short between D+ to D–. This allows the creation of DCP "wall warts," which feature a USB mini- or micro-receptacle instead of a permanently attached wire with a barrel or customized connector. Such adapters allow any USB cable to be used for charging.

The main task for a device connecting to a USB port is to know how much power can be drawn from the USB port. Overdrawing power from a current-limited USB port could result in a USB crash. The USB often will not restart until the device is unplugged and reconnected. A portable design has choices about how to manage port detection. It can be compliant with BC1.2, compliant only with USB 2.0, or noncompliant. If fully compliant with BC1.2, it must be able to sense and limit input current for all USB source types, including legacy USB 1.0 and 2.0 ports. If compliant with 2.0, it will charge from SDPs after enumeration, but may not recognize CDPs and DCPs. If it cannot recognize a CDP, it can still charge and remain compliant but only after enumeration, in the same way that it would with an SDP. Other partially compliant and noncompliant charging schemes are discussed later.

A device can implement port detection using its own software, or it can use a charger or interface IC that detects by interacting with the USB D+ and D– data lines without relying on system resources. The design's partitioning of these roles depends on the system architecture. For example, a device that uses a microcontroller, or a dedicated IC, to manage power may prefer to use that

IC for port detection. Since the device already can communicate with the host over the USB connection, it can make charging choices based on the results of enumeration and configuration. A different device might not be designed to communicate with a USB or might not want to devote system software to manage USB charging. It just wants to use available USB ports as a power source. This approach can be used to avoid complexity or in response to worries that software bugs might cause incorrect charging.

2.3.7   Port Detecting and Self-Enumerating Charger

The MAX8895 from Maxim Integrated Circuits determines how effectively to use available input power without relying on the system to determine the power source output power capability. The charger automatically determines the adapter type and can distinguish between:

- *DCP:* 500 mA to 1.5A.
- *CDP (host or hub):* to 900 mA for high speed; to 1.5A for low and fast speed.
- *Low-power SDP (host or hub):* 100 mA.
- *High-power SDP (host or hub):* 500 mA.

The available current can be used by the battery or the system, or it can be split between them. A built-in suspend timer automatically triggers a suspension when no bus traffic is detected for 10 ms. In addition to automatically optimizing current from USB and adapter sources, it also deftly handles switchover from adapter and USB power to battery power. It also allows the system to use all available input power when necessary. This enables immediate operation with a dead or missing battery when power is applied. All power-steering MOSFETs are integrated, and no external diodes are needed. Die temperature is kept low by a thermal regulation loop that reduces charge current during temperature extremes.

2.4   Battery Charger and System Interactions

Figure 2.19 shows the most commonly used battery charging and system power architecture, where the system is directly connected to the battery. The output of the charger is first to charge the battery. In addition, the battery charge output provides power to the system, which makes this architecture simple and low cost. Many linear chargers such as the LTC4059 from Linear Technologies, ADP3820 from Analog Devices, ISL6293 from Intersil Corporation, and

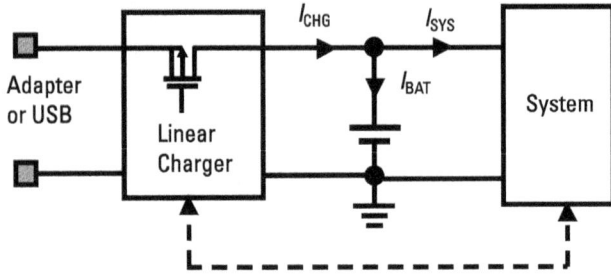

Figure 2.19   Battery charging and system power architecture block diagram.

bq24010 from Texas Instruments are available. However, connecting a system load to the battery can cause various issues such as longer battery charge times, charge termination, and false safety timer warnings. In this configuration, the charger output current I_{CHG} is not dedicated to charging the battery, but is instead shared between the system and the battery. Current I_{CHG} is the current that the charger can control. The battery charger makes charging decisions based on this current, and is not aware that the system steals some of the charge current. Therefore, the charger is not able to directly monitor and control the effective battery charge current I_{BAT} into the battery.

During the precharge phase, the precharge current is typically 10% of the fast-charging current for a deeply discharged battery (less than 3.0V). The system load I_{SYS} steals away some portion of this current and the effective charge current becomes even smaller. For example, the precharge current of the charger output is 100 mA while the system draws 60 mA. As a result, the effective precharge current into the battery is only 40 mA. This not only increases the battery precharge time, but also may cause a precharge timer false expiration if the battery voltage does not rise to 3V within the precharge timer period. This may provide a false precharge safety timer warning because of not enough precharge current, but not because of a defective battery. It is even possible for the system current to be larger than the precharge current, in which case the battery would be discharged instead of being charged even with the connected power source. The discharge protection MOSFET could be turned off in the battery pack when the battery cell reaches the undervoltage protection threshold and the system will be permanently locked off. To solve this issue, the system has to be either in shutdown mode or low-power standby mode so that the precharge current is dedicated to charging the battery to above 3.0V within the precharge safety timer period. This is the main reason why you may not be able to turn on a cell phone or even make a phone call when the adapter is connected to a deeply discharged battery. Similarly, once the battery enters the fast-charge phase, the system load still steals some charge current from the charge output, which increases the battery charge time and may result in fast safety time false expiration.

Another possible issue is charge termination. The charger will monitor the battery charge current so it can terminate the charging process when its output current falls below the charge current termination threshold. As we discussed, the charger does not detect the effective charge current; instead, it monitors the total current of the system and battery. If the system has a constant DC current higher than the charge termination current threshold, then the charger will never be able to detect the charge termination, which results in the false safety timer expiration. How can we solve these application-related issues and improve system performance? Various dynamic power path management techniques that monitor the real effective battery charge current are discussed next to solve these problems.

2.5 Dynamic Power Management Battery Charger

The issues mentioned in the preceding section are caused by the interaction between the charger and the system. The charger cannot detect the real effective charge current for making the right charge decisions. To operate the system while charging a deeply discharged battery, a power path management technique should be used to eliminate the system and charger interaction. Powering the system and charging the battery should have independent power paths.

2.5.1 System Bus Voltage-Based Dynamic Power Path Management (DPPM) Charger

Figure 2.20 shows a simplified linear battery charger block diagram with a power path management. MOSFET Q1 is used to preregulate the system bus voltage V_{OUT} in LDO mode or fully turned on for obtaining the maximum input power. This establishes a direct power path from the input to the system for

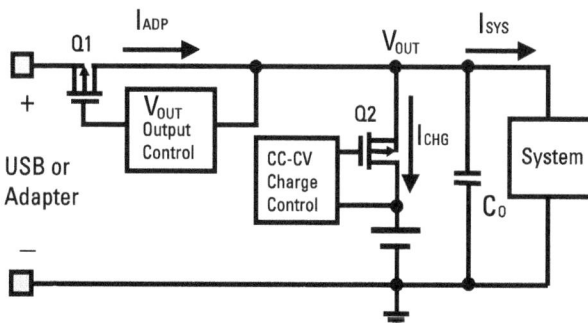

Figure 2.20 Power path management battery charger block diagram.

providing power to the system. MOSFET Q2 is dedicated to fully controlling the battery charging in a linear charger fashion. The battery charger controller and MOSFET Q2 are composed of a linear battery charger, and the system bus voltage is used as the input voltage of a linear battery charger, which should be always higher than the battery voltage and also should be high enough to power the system. Because there are two independent power paths for powering the system and charging the battery, battery and system interference no longer exists. This power architecture establishes two separate paths for system and battery charging, called power path management (PPM). The dedicated battery charging path is able to not only power the system even without a battery or a deeply discharged battery, but also can completely eliminate the false safety timer expiration and charge termination issues as well. It allows the system to operate while charging a deeply discharged battery since the system bus voltage is regulated to a set value such as 4.4V, for example, or adapter voltage through MOSFET Q1 with or without a battery.

Figure 2.21 shows the waveforms for the DPPM operating principle. Since the system current is usually pulsating, its peak system current could be higher than the input peak current limit. The capacitor C_o connected in the system bus starts to discharge and system bus voltage drops at t_1 when the current required from the system and battery charger is higher than the amount

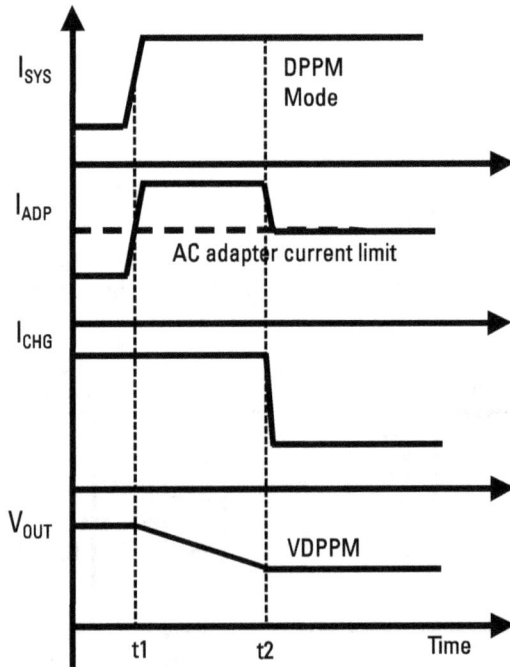

Figure 2.21 DPPM charger waveforms.

of input current available from the AC adapter or USB. Once the system bus voltage falls to keep the preset DPPM threshold at t_2, the battery charge control regulation loop is going to regulate the system bus voltage by reducing the battery charging current so that the total current demand from the system and battery charger is equal to the maximum current available from the adapter to prevent the system bus voltage from dropping further. The DPPM control loop tries to reach a steady-state condition where the system gets its needed current and the battery is charged with the remaining current. This maximizes the use of the power available from the adapter or USB. Most system loads are very dynamic with a high pulsating current. The adapter will be overdesigned if its power rating is based on the maximum peak power from the system and battery charger since the average power from the system is much smaller than its peak maximum power.

The DPPM control technique allows the system to use a smaller power rating and cheaper AC adapter while supplying system power and charging the battery simultaneously. The system bus voltage could be fluctuated between its maximum input voltage or preregulated voltage and DPPM threshold voltage. Such fluctuated voltage may cause audible noise if the frequency is below the audible frequency of 20 kHz.

Figure 2.22 shows a DPPM Li-ion battery charger example with the following design specifications: 800-mA fast-charge current and a 5-hr safety timer. Whenever the charge current is reduced due to either active thermal regulation or active DPPM operation mode, the safety timer is automatically

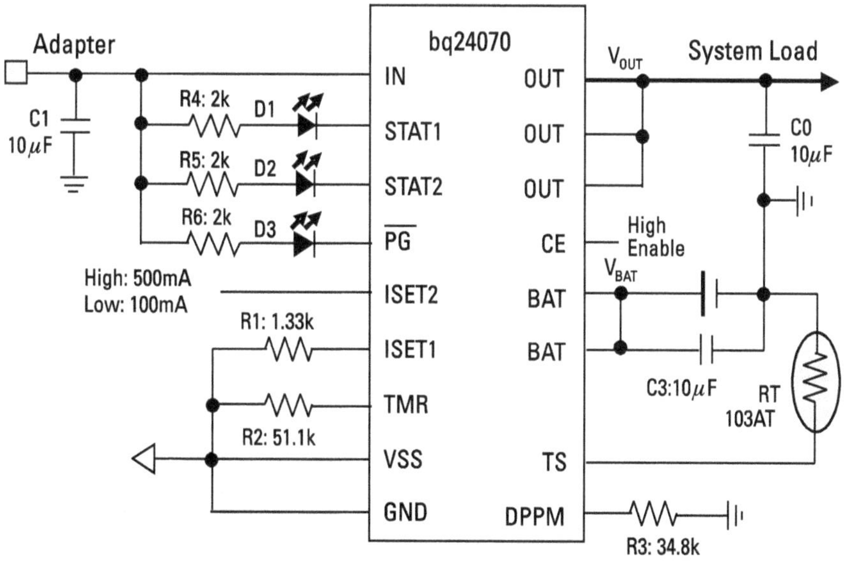

Figure 2.22 Dynamic power path management battery charger.

adjusted to increase the timer value, which avoids unexpected false safety timer expiration. In addition, the charge termination is also disabled to prevent false charge termination when either the DPPM or thermal regulation loop is active.

2.5.2   Input Current-Based Dynamic Power Management (DPM) Linear Charger

The input power source (e.g., an adapter) has an output current limit. Its output voltage may be crashed when the current drawn by the system exceeds the adapter current limit. Two options are available for avoiding an adapter voltage crash. One is to increase the adapter power rating to match the peak system power and maximum charge power demand. This will, however, increase the adapter size and cost. The second option is for the charger control system to monitor the input current and automatically regulate the battery charging system so that the total current required by the charging system is below the input power source current limit. In other words, the charger system gives higher priority to powering the system and the remaining power will be used to charge the battery. Figure 2.23 shows a block diagram for an input current-based DPM linear charger. MOSFET Q2 is dedicated to regulating the battery charge voltage and current in a linear fashion, while MOSFET Q1 is used to regulate the system bus voltage. Figure 2.24 shows its operating waveforms. When the system current increases at t_1, the input current I_{IN} immediately exceeds the input current limit. The input current is fed into the error amplifier A2, compares the input current limit threshold, and then regulates the on-resistance of MOSFET Q2 for reducing the charge current. The input current reaches it reference input current I_{REF} at t_2 when the charge current drops to I_{CHG2} at t_2. When the peak system current is removed at t_3, the input current is below the input current

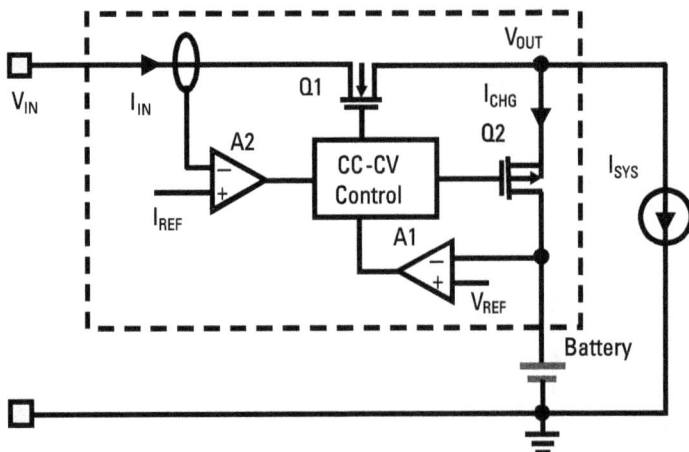

Figure 2.23   Input current–based DPM linear charger block diagram.

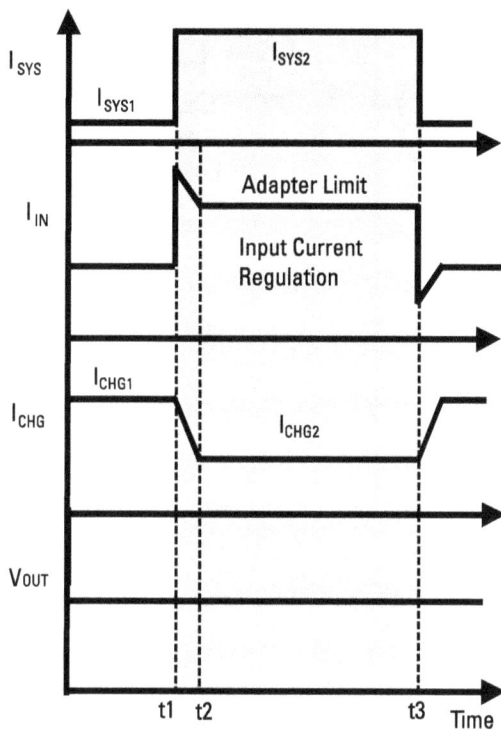

Figure 2.24 Operating waveforms of the input current-based DPM linear charger.

limit and the battery charge current is regulated at its fast-charge current for maximizing the charging speed. The main advantage is that the system bus voltage is DC constant without audible noise.

Figure 2.25 shows a typical application circuit that uses the input current-based DPM linear charger LTC4066 from Linear Technologies. It fully supports USB 100 mA/500 mA charging, and has the charge current output for both charging and discharging, which can be used as an ADC input for fuel gauging. An ideal diode function between the system output and battery provides power from the battery when output/load current exceeds the input current limit or when input power is removed. Powering the load through the ideal diode instead of connecting the load directly to the battery allows a fully charged battery to remain fully charged until external power is removed. Once external power is removed, the output drops until the ideal diode is forward biased. The forward-biased ideal diode will then provide the output power to the load from the battery. Furthermore, powering switching regulator loads from the OUT pin rather than directly from the battery results in shorter battery charging times. This is due to the fact that switching regulators typically require constant input power. When this power is drawn from the OUT pin voltage,

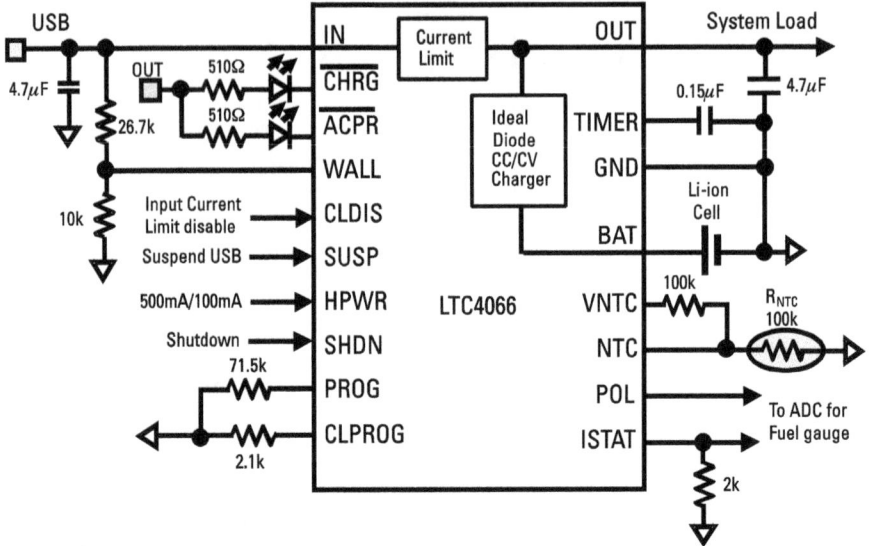

Figure 2.25 Input current-based DPM linear charger with power path and ideal diode operation.

which is higher than the battery voltage, the current consumed by the switching regulator is lower, leaving more current available to charge the battery.

Figure 2.26 shows another application circuit of the input current-based DPM linear charger with a 1.3A input current limit, 800-mA fast-charge current, 6.25-hr safety timer, and 0° to 50°C charge temperature window. The ILIM pin is used to set the maximum input current limit, and ISET is employed to set the fast-charge current. The battery charge safety timer is set by TMR. To support the USB charging, EN1 and EN2 pins will set the 100- or 500-mA input current limit for supporting USB 2.0.

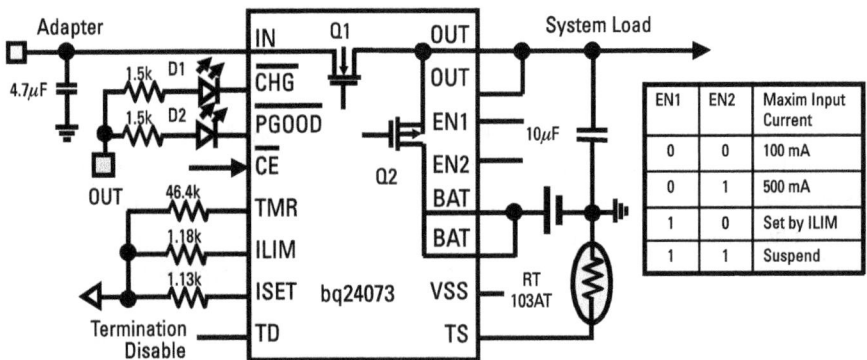

Figure 2.26 Input current-based DPM linear charger application circuit.

A few input current DPM chargers are also available such as the LTC4055 from Linear Technologies and the MAX8677A from Maxim Integrated Circuits.

2.5.3 Switch-Mode DPM Battery Charger with Power Source Selector

As mentioned in the discussion about the switch-mode charger being more efficient than a linear charger, many applications (e.g., laptops, tablets, and industrial applications) require that the battery be charged and the system powered simultaneously. Figure 2.27 shows a switch-mode DPM battery charger with power source selector. MOSFETs Q1 and Q2 are composed of a synchronous switching buck converter and turn on and off alternatively. Inductor L and output capacitor C_o are used as a low-pass filter to filter out all switching ripple for getting the DC output. Current-sense resistor R_{SNS} is used for sensing the charge current, and its current information is fed into the charge current regulation loop to make sure that the charge current does not exceed the reference current. The current-sense resistor R_{AC} is employed to monitor the input current fed into the PWM control loop for regulating the duty cycle of MOSFET Q1 so that total current drawn by the charger and system does not exceed the adapter current limit.

The system can be powered by either an input power source from the adapter or the battery. The input power source usually provides the power to the system while the remaining power is used to charge the battery. Whenever the input power source is not connected, the battery is used as a power source

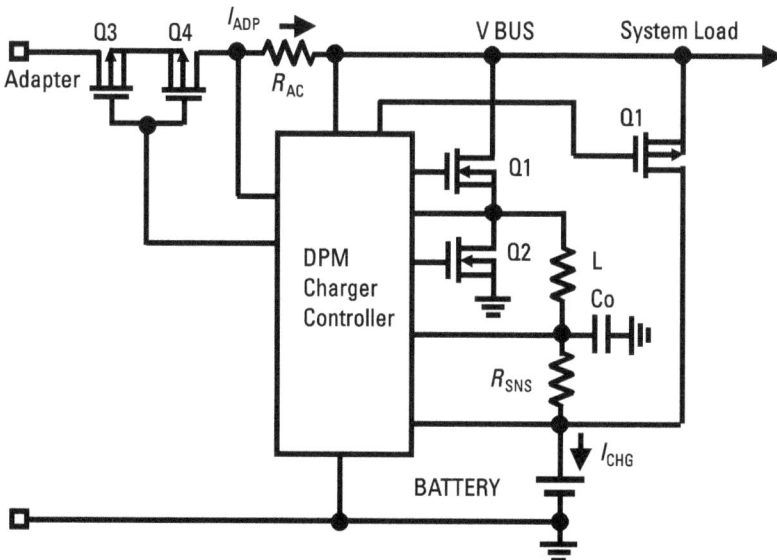

Figure 2.27 Switch-mode DPM battery charger with power source selector.

to power the system. MOSFETs Q3, Q4, and Q5 are used as power selector switchers. When the input power source is connected, MOSFETs Q3 and Q4 are turned on, while MOSFET Q5 is off. Battery MOSFET Q5 is turned on and MOSFETs Q3 and Q4 are off while the input power source is removed. Why are the two back-to-back MOSFETs Q3 and Q4 required at the input? The MOSFET usually has an antiparallel body diode. MOSFET Q3 blocks a battery leakage path from the battery to the input through the body diode of Q4. MOSFET Q4 is employed to limit the input inrush current by slowly turning it on as the amount of capacitors is connected in the system bus. To prevent two power sources from shoot-through, there is dead time between Q5, Q3, and Q4 so that they cannot be turned on at the same time.

The DPM function is similar to that of a linear DPM charger. When the system current increases, the input current I_{ADP} increases accordingly. If the input current reaches its preset current limit, its input current regulation loop will be activated and maintain the input current as a constant by reducing the PWM duty cycle of the MOSFET Q1 for reducing the effective charge current. The charge controller gives higher priority to powering the system while the remaining power from the adapter is used to charge the battery.

Figure 2.28 shows the switch-mode Li-ion battery charger with power source selector for supporting a one- to six-cell series battery pack, which typically includes an AFE for achieving all kinds of overcurrent protections and

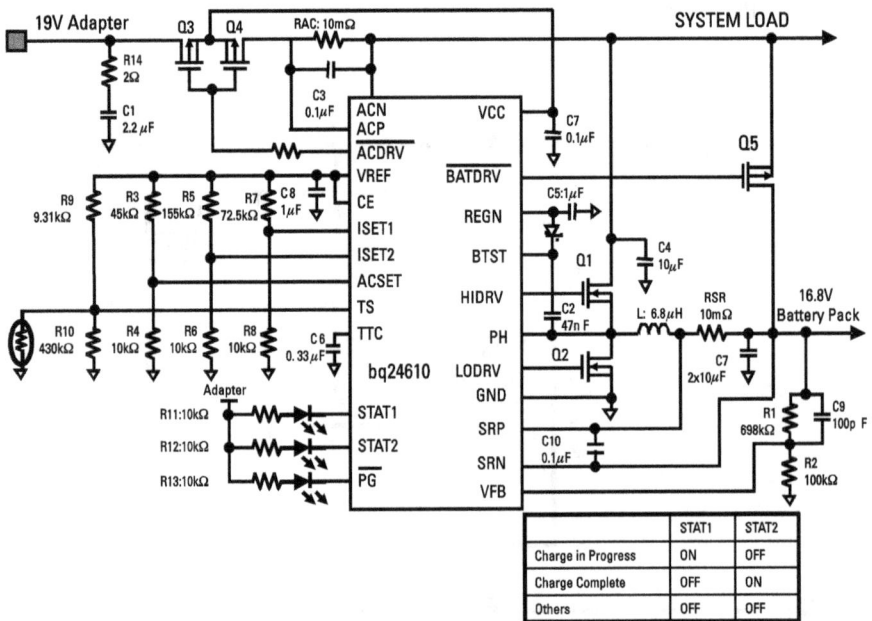

Figure 2.28   DPM switch-mode charger example with power source selector

overvoltage protections, or a second-level protector, and may include a fuel gauge. (The battery pack electronics safety design will be discussed in Chapter 3.)

The charge specifications are as follows:

Charge voltage: 16.8V for a four-cell series battery pack.

Fast-charge current: 2A.

Precharge current: 200 mA.

AC adapter current limit: 3A.

Safety timer: 3 hours.

Charge temperature qualification window: 0° to 45°C.

The power source selector is used to choose either the adapter or battery for powering the system. Resistor R14 and capacitor C1 are used as an input RC filter for damping the input voltage spike due to the resonance between the adapter cable inductance and input capacitance when the adapter is plugged. Six-volt gate drive voltage from the REGN LDO output is used effectively for turning on and off the MOSFET Q1 and Q2 in the synchronous buck converter. Why does it use 6V gate drive voltage instead of 5V? The on-resistance of the MOSFET is inversely proportional to the gate drive voltage. The higher the gate drive voltage, the lower the on-resistance. Six-volt gate drive voltage can reduce the MOSFET on-resistance by 10% smaller than that of 5V gate drive voltage. The power loss saving from the on-resistance reduction is much more than the loss increase from the gate driver. The battery regulation voltage can be set through the resistor divider R1 and R2 for achieving any battery charge voltage, ranging from a one- to six-cell series battery pack. The fast-charge current can be set through a resistor divider of R7 and R8 at the ISET pin, while the precharge current can be set through the resistor divider of R5 and R6 at the ISET2 pin. The maximum input current limit can be set by the voltage at the ACSET pin. The battery safety timer is determined by the capacitance connected at the TTC pin, while the battery charge temperature qualification is achieved by the voltage at the TS pin. The battery charge status such as "charge in progress" and "charge done" can be determined via the status indicator outputs of STAT1 and STAT2. This is a stand-alone charger because the battery charger does not rely on the charger controller for achieving the battery charge function from the precharge to fast-charge transition, charge current and charge voltage regulation, battery charge temperature qualification, safety timer, and battery charge status outputs. It does not employ an embedded microprocessor to achieve any charge function. The main advantage is that it is a simple, low-cost system design. However, multiple hardware designs are required for different battery packs with different chemistries when the charge voltage and current are different.

A few stand-alone switch-mode chargers are available such as the bq24610 from Texas Instruments, bq24610, the MAX1737 from Maxim Integrated Circuits, and the LTC4007 from Linear Technologies. Table 2.1 compares their features and performance.

2.5.4   Narrow Voltage Direct Current (NVDC) DPM Battery Charger

In the traditional DPM switch-mode charger, the system bus voltage has a wide voltage range from minimum battery voltage to the adapter input voltage. If a 12V input power source is used to charge a single-cell battery, the system voltage will vary from the 3.0V of minimum battery voltage to 12V adapter voltage. It is difficult to achieve a high-efficiency downstream DC-DC converter with such a wide range of system bus voltage. Furthermore, we may have to insert an intermediate DC-DC converter like buck-boost converter if the input voltage of the downstream DC-DC converters cannot exceed 6V maximum voltage. Therefore, it is very desirable to have a narrow system bus voltage for

Table 2.1

Input Current DPM Charger Comparison

Function	bq24610	MAX1737	LTC4007
Charger Control Architecture	Synchronous Buck Constant Frequency N-MOSFETs; 600kHz	Synchronous Buck Constant Frequency N-MOSFETs; 300kHz	Synchronous Buck Constant off-time P/N-MOSFET; 300kHz
Compensator	Internal	External (CCV/CCI/CCS)	External (ITH)
Power Path Control	Yes	No	Only input FET
Input Current DPM	Yes	Yes	Yes
Cells	1-6 Cell Series	1 to 4-Cell Series	3 or 4-Cell
Battery Voltage	Programmable FB	4.2V +/-5%/Cell	4.1/4.2 (CHEM)
Charge Current	Programmable SET1	Programmable ISETOUT	Programmable PROG
Accuracy	Voltage: 1%, charge current:3%, input current 3%	Charge voltage 1%; charge current: 7.5%; input current 5%	Charge Voltage: 1%, charge current: 5%; input current: 7%
Pre-charge current	Programmable ISET1	Fixed C/20	Fixed 10%
Timers/Termination	TTC	TIMER1, TIMER2	1-3h (15%) max (RT)
Re-charge (V/cell)	100mV/Cell	3.81 / 3.9	3.81 / 3.9
Status/Interface	STAT1, STAT2	/FAULT, /FASTCHG, CC mode, CV mode	/LOBAT, /FAULT, /CHG, /FLAG,
Temperature Sense	Programmable TS	Programmable THM	Programmable NTC
Package	4mmx4mm QFN-24	6mmx8mm QSOP-28	6mmx8mm SSOP-24 4mmx5mm QFN-24

achieving the highest system efficiency. The main difference between the DPM charger shown in Figure 2.27 and a NVDC charger is that with a DPM charger the rest of the system sees the adapter voltage (AC present) or the battery voltage (AC absent), whereas with a NVDC charger, the system always sees the battery voltage (or close to it), regardless.

Figure 2.29 shows a NVDC switch-mode battery charger. The battery charger is not only used as a charger, but also a DC-DC stepdown converter. MOSFETs Q1 and Q2, inductor L, and output capacitor C_o are composed of a synchronous buck converter to get the desirable system bus voltage. Similarly, current-sense resistors R_{AC} and R_{CHG} are used to sense the input current and charge current, respectively. MOSFET Q3 is employed to block the battery leakage path from the battery to the input. For a deeply discharged battery, the DC-DC converter provides a minimum system operating voltage, while MOS-FET Q4 operates in LDO mode to provide the precharge current. Once the battery voltage reaches the minimum system bus voltage, the MOSFET Q4 will be completely turned on and the DC-DC converter regulates the fast-charge current and charge voltage while the battery voltage reaches the minimum battery voltage. Therefore, the system bus voltage ranges from the minimum battery voltage to the maximum battery voltage, which allows the system to achieve the highest efficiency. The MOSFET Q4 operates as a linear charger only in the precharge mode, but not in the fast-charge mode. The inductor and DC-DC power stage have to carry both the charge current and the system power, which needs high current ratings. However, for the traditional DPM battery charger, the DC-DC converter deals only with the battery charge current; the system

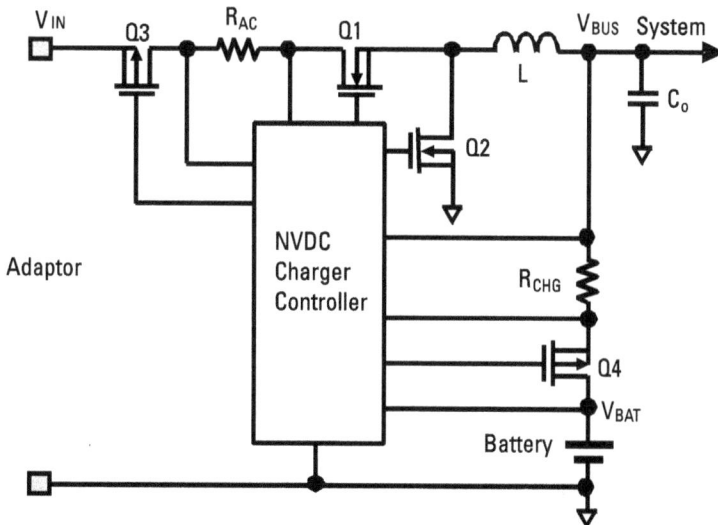

Figure 2.29 Switch-mode NVDC battery charger.

power is not processed by the charger. Therefore, a NVDC charger could be suitable for lower power applications.

A few NVDC charger controllers are available, such as the ISL9519 from Intersil Corporation, the bq24190 and bq24715 from Texas Instruments, the MAX8903 from Maxim Integrated Circuits, and the LTC4099 from Linear Technologies.

2.5.5   Battery Charging System Topology Comparisons

Each of the above-mentioned battery charging system topologies has its advantages and limitations, depending on the application and design considerations that affect performance, such as efficiency, solution size, and cost. Table 2.2 gives a comparison of the various topologies.

2.6   Battery Charger Design Examples in End Equipment

2.6.1   Tablet Charger Design Example

Most of today's tablets use a single-cell battery with from 3- to 10-Ah capacity. Among them, 6–7-Ah and 3–5-Ah battery packs are popular for 10-inch tablets and 7-inch tablets, respectively. Based on a typical 0.5C to 0.7C charge rate, a 4A charger is preferred for 10-inch tablets and a 2.5A charger for 7-inch tablets. The main power sources for powering the tablet include USB 2.0 and 3.0 ports and AC adapters, which could share the same connector. The system operating

Table 2.2
Battery Charging System Topology Comparisons

Function	Linear Charger	DPM Linear Charger	DPM Switch-Mode Charger	NVDC Charger
Size	Small	Moderate	Larger	Larger
Complexity	Simple	Moderate	Complex	More complex
Solution Cost	Low	Moderate	Higher	Higher
Power Source Control	No	Yes	Yes	Yes
System Instant-On	No	Yes	Yes	Yes
Charge Efficiency	Low, V_{BAT}/V_{IN}	Low, V_{BAT}/V_{IN}	Excellent, >90%	Excellent, >90%
Power Dissipation	High	High	Low	Low
Charging Deep While Operating the System	No	Yes	Yes	Yes
System Bus Boltage	Narrow, battery voltage	Narrow, battery voltage	Wide (input voltage and battery voltage)	Narrow, battery voltage
EMI	No	No	Potential	Potential

voltage of 3.5V is required for keeping the system operating even with a deeply discharged battery. The tablet charger design specifications are as follows:

- Battery pack capacity: single-cell 6,000 mAh.
- Charge voltage: 4.2V.
- Fast-charge current: 4A.
- Precharge current: ≤300 mA.
- Precharge to fast-charge voltage threshold: 3.0V.
- Battery temperature qualification range: 0° to 45°C.
- Minimum system operating voltage: 3.4V.
- USB OTG output current: up to 1A.
- Adapter: 12V at 2A.
- USB 2.0 and USB 3.0 support.

Figure 2.30 shows a 4A I2C controlled synchronous switching NVDC battery charger for meeting the above design specifications. Many of the charge parameters are controlled by the I2C registers, which are very flexible. These I2C controlled registers include input power source current limit, precharge current, fast-charge current, charge voltage, safety timer, minimum system operating voltage, charge termination, watch-dog timer, and programmable thermal temperature regulation. Fast charging can be achieved by setting the maximum charge current of 4A. But the challenge is the heat generated by the power consumption from the switching loss and conduction loss of the four integrated power MOSFETs. To minimize the losses, the on-resistance of the MOSFETs should be designed in such a way that the temperature rise should be limited to 50°C in a practical design given that the ambient temperature could reach as high as 50°C. This charger has four programmable thermal regulation temperatures of 60°, 80°, 100°, and 120°C by reducing the charge current when the IC temperature reaches the thermal regulation temperature. The designers can achieve the best thermal performance by setting 60°C while maximizing the charge speed by setting 120°C. Another unique fast-charging algorithm of IR compensation is used for extending the constant current charge mode and shorting the constant voltage mode operation by increasing the charge voltage for compensating the voltage drop across the on-resistance of the two back-to-back protection MOSFETs and current-sense resistance for accurate fuel gauging in the battery pack. Such IR compensation can reduce the battery charge time by about 10% to 20%.

How do we achieve the maximum battery run-time when the adapter is not connected? For a single-cell operating system, the typical minimum system

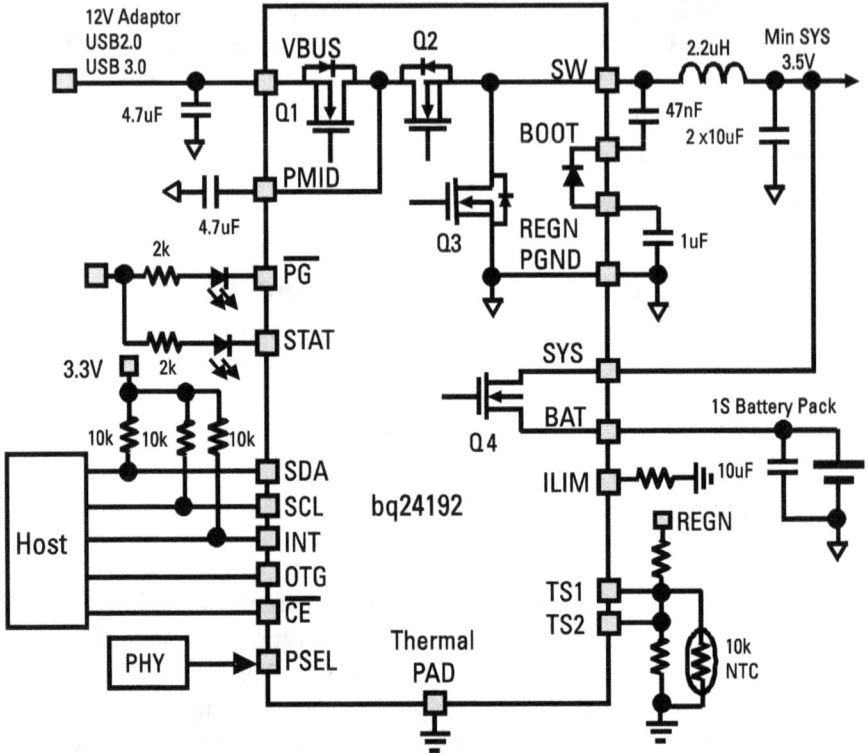

Figure 2.30 A 4A I2C single-cell NVDC battery charger for a 10-inch tablet.

voltage is around 3.4V in order to reach the 3.3V output usually required by a system. If the on-resistance of MOSFET Q4 is 50 mΩ and the battery discharge current is 3A, the battery cut-off voltage is 3.55V. This means that all battery energy under 3.55V will not be used and such battery energy could reach as high as 15% of the battery capacity not used by the system. Therefore, the on-resistance of MOSFET Q4 is very critical and 12 mΩ of on-resistance is used in this design. How do we determine if the power source is a USB port or adapter? A PHY device is used to detect whether the input power is from a USB port or adapter and then the input current limit is set accordingly.

Tablet chargers are available from Linear Technologies (LTC4155), Texas Instruments (bq24192), and Maxim Integrated Circuits (MAX8903). Comparisons of these chargers are shown in Table 2.3.

2.6.2 Notebook and Ultrabook Battery Charger Design Example

Notebooks and ultrabooks have been becoming one of the most important devices in daily life. Ultrabooks require a lower power CPU than do notebooks

Table 2.3
Tablet Charger Comparison

Function	LTC4155	bq24192	MAX8903
Input Voltage Range	Adapter and USB: 4.35V–5.5V,	Adapter: 3.9V–17V USB: 3.9V to 6.5V	AC: 4.15V–16V USB: 4.1V–6.3V
Maximum Input Current	3A	3A	2A
Charge Voltage	4.05V–4.20V, 50mV/step	3.6V–4.5V, 16mV/step	Fixed 4.2V
Maximum Charge Current	3.5A	4.5A	2A
Charge Efficiency	87.5%@2A, 80%@3A	94% @2A; 90%@4A	86%@2A
Input Voltage DPM for Supporting All Adapters	No	Yes	No
USB Authentication (D+, D−)	No	Yes	No
USB Support	USB 2.0	USB 2.0 and 3.0	USB 1.0
USB OTG	Yes, 500mA	Yes; 1.3A	No
OTG Efficiency	Not specified	92%@5V/1A	Not applied
Battery Discharge FET Rdson	External P-MOSFET	12mΩ	50mΩ
Battery Fast Charging Algorithm	No	Yes	No
Switching Frequency	2.25MHz	1.5MHz	Up to 4MHz (1.7MHz @V_{in}=5V, V_{bat}=4V)
Control Interface	I2C	I2C	Standalone
Package	28-pin 4mmx5mm QFN	24-pin 4mmx4 mm QFN	28-pin 4mmx4mm QFN

and they are less than 20 mm thick. In traditional mobile computer systems, an AC adapter provides power to mobile computers. If power demand from the system is lower than the adapter power, the remaining power is used to charge the battery. When an AC adapter is not available, the battery provides power to the system by turning on switch S1 as shown in Figure 2.31. The adapter can be used to power the system and charge the battery simultaneously, which may lead to use of a high-power rating adapter, increasing both the size and cost without active control. A dynamic power management technique typically is used to accurately monitor the total power drawn from the adapter, which gives high priority to powering the system.

Once the adapter power limit is reached, the DPM control system regulates the input current (power) by reducing the charge current while providing the power directly from the adapter to the system without power conversion for optimum efficiency. With the heaviest system load, the adapter power is completely used to power the system without charging the battery at all. Therefore, the main design criteria are to make sure that the adapter power is high enough to support peak CPU power and other system power.

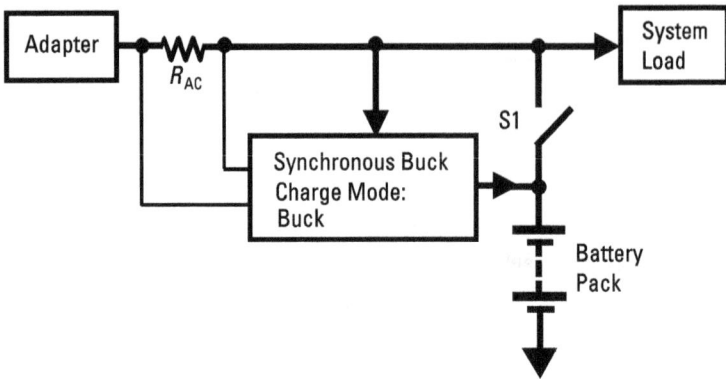

Figure 2.31   Adapter and battery charger system block diagram.

The demand to improve a system's performance, such as processing complicated tasks faster with multiple CPU cores and enhanced graphics processor units (GPUs), is continuous. Turbo boost technology, developed by Intel in the Sandy Bridge processor, allows processors to burst their power above the thermal design power (TDP) for a short time period in the range from a few tens of milliseconds to tens of seconds. However, an AC adapter is designed to provide power just above the amount demanded by the processors and platform at a TDP level considering the design tolerance. When a charger system detects that the adapter input power has reached its power rating after its charge current is reduced to zero, the simplest way to avoid crashing the AC adapter is to achieve CPU throttling by reducing the CPU frequency, which compromises system performance. How can we continuously improve system performance by operating the CPU faster at the above TDP power level for a short time period while not crashing the adapter and not increasing the adapter power rating?

A hybrid power boost charger for notebooks and ultrabooks could an effective solution. When the total power required by the system load and battery charger reaches its adapter power limit, the battery charge current starts to decrease through the DPM regulation loop. The battery charger stops charging, and its charge current decreases to zero when the system load alone reaches the AC adapter power limit. As the system continues to increase its load in the CPU's turbo mode, the system load exceeds the AC adapter limit for a short time period, which could result in crashing the AC adapter. One possible solution is to increase the AC adapter power rating to meet the maximum peak system power, but at the expense of additional adapter cost. During the CPU's turbo mode, the battery charger, which is usually a synchronous buck converter, is idle as no remaining power is available to charge the battery. As we know, the synchronous buck is actually a bidirectional DC-DC converter. It can operate in both buck mode and boost mode, depending on the operating conditions. If the battery has enough capacity, the battery charger can operate in boost mode to provide additional power to the system in addition to the power from the AC adapter. Figure 2.32 shows the hybrid power boost battery discharger block diagram for supporting the CPU turbo mode.

When and how does the battery charger start to transition from buck charge mode to boost discharge mode? The system can enter the CPU turbo mode at any time, and it is usually too late to inform the charger to operate the charger transition from charge mode to discharge boost mode through an SMBus. The charger should automatically detect which operating mode is needed. Also critical is to design the charger system so that is achieves a fast-mode transition from the buck charge mode to the boost discharge mode, and vice versa. A DC-DC converter needs a soft-start time (from a few hundred microseconds to a few milliseconds) to minimize the inrush current. During this mode transition time period from buck charge mode to boost discharge mode, the adapter should have a strong overloading capability to support the whole system peak power before the charger transitions into boost discharger mode. Most of the AC adapters currently available can hold their output voltage over a few milliseconds.

Figure 2.33 shows the hybrid power battery charger application circuit for supporting a CPU's turbo mode. The R_{AC} current-sense resistor is used to detect the AC adapter current for DPM function, and determine whether the battery charger operates in buck charge mode or in boost discharge mode. Current-sense resistor R7 is used to sense the battery charge current programmed through the SMBus from the host based on the battery conditions. The total power drawn by both charger and system can be monitored through the I_{OUT} output, which is 20 times the voltage drop across the sense resistor RAC for

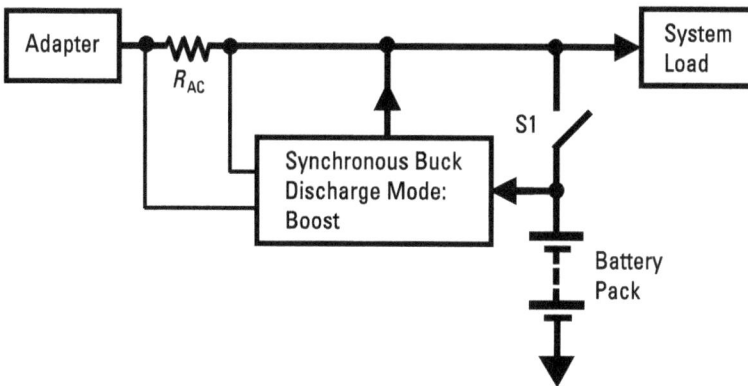

Figure 2.32   Hybrid power boost battery block diagram in the CPU turbo mode.

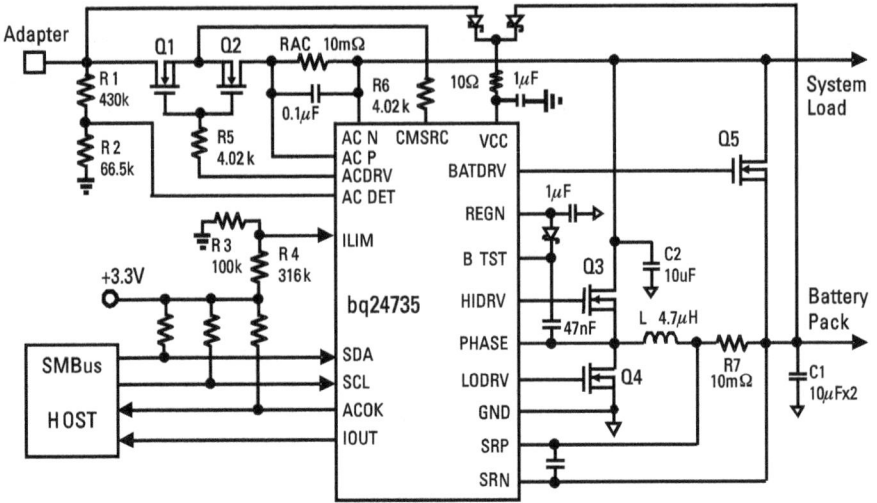

Figure 2.33 Hybrid power boost battery charger application circuit.

achieving the CPU throttling, if needed. The turbo-boost battery discharge mode can be enabled or disabled through SMBus control registers based on the battery state of charge and battery temperature conditions.

To achieve a small form factor for a notebook computer like an ultrabook driven by Intel, the switching frequency can be programmed at 615, 750, or 885 kHz. This minimizes the inductor size and the number of output capacitors. To further reduce the external components, the loop compensators for the charge current, charge voltage, and input current regulation are fully integrated in the charger controller chip. The power source selector MOSFET controller is also integrated in the charger. Furthermore, the charge system uses all N-channel MOSFETs for cost reduction, instead of P-channel power MOSFETs in the traditional charge solution. The benefit of this hybrid power battery charger is that the same battery charger system can be used without changing the bill of materials. System designers can do a quick system performance evaluation without any additional hardware design effort.

Figure 2.34 shows the switching waveforms about the mode transition between the buck charge mode and boost discharge mode. When the input current reaches the maximum input current limit due to the system load increase, the battery charger stops charging. The battery then transitions into boost mode to provide additional power to the system. If the battery is removed or the battery's remaining capacity is not high enough, CPU throttling is necessary to avoid an adapter crash.

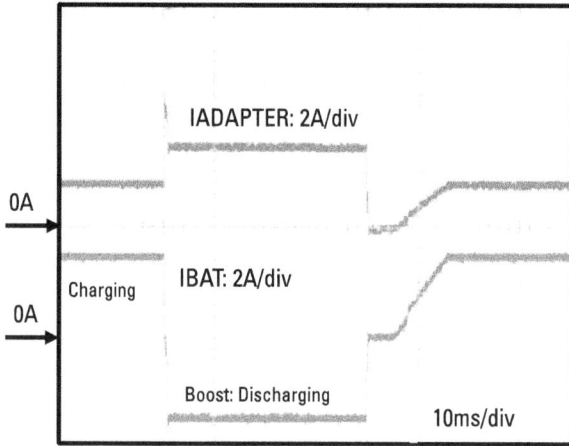

Figure 2.34   Waveforms between buck charger mode and boost discharger mode.

2.7   LiFePO$_4$ Battery Charger

The existing Li-ion batteries with lithium cobalt oxide and lithium manganese oxide as the cathode materials suffer from a low discharge rate, safety concerns, and short cycle life. When the cell temperature increases to 175°C with a 4.3V charge cell voltage, the battery enters the thermal runaway mode and could cause an explosion. To solve these problems, Dr. John Goodenough and his team at the University of Texas patented the LiFePO$_4$ as a potential cathode in 1996. It is a very stable material due to its covalent P–O bonding, which stabilizes the fully charged cathode. Scientists have developed the LiFePO$_4$ battery using lithium iron phosphate as a cathode to alleviate safety concerns in the battery for electric vehicles and power tool applications.

Figure 2.35 shows the LiFePO$_4$ battery's discharge characteristics under different discharge currents from company A123. Unlike the LiMnO$_2$ cathode-based Li-ion battery, its discharge curve is fairly flat and its voltage is almost constant. It has a very strong discharge capability with high discharge rates, and is suitable for power tools and electric vehicle applications. The LiFePO$_4$ cycle life could be longer than 1,000 cycles at room temperature (25°C/77°F), and can reach 1,000 cycles even at 60°C/140°F compared with the average life of 300 to 500 cycles for a Li-ion battery. However, the main drawback is its energy density, which is only about 50% of the standard Li-ion battery.

The LiFePO$_4$ battery charging profile is similar to that of the traditional Li-ion battery. It also uses constant current (CC) and constant voltage (CV) and consists of three charging phases: precharge, fast-charge CC, and CV. Fast-charge current is applied to charge the battery quickly. Its charging rate can be as high as 10C, which is much higher than that for a traditional Li-ion battery

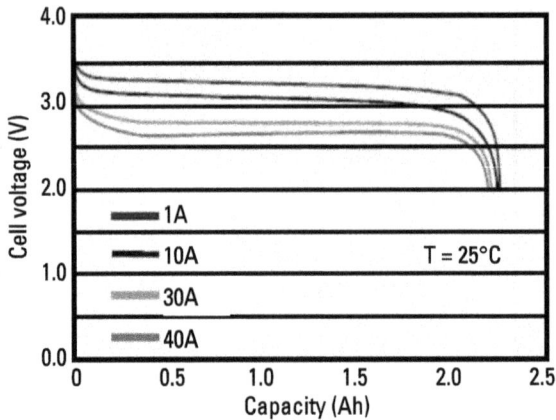

Figure 2.35   LiFePO$_4$ battery discharge characteristics.

without additional degradation. The charger enters to the CV mode when the battery reaches a voltage regulation limit (typically 3.6 V/cell). During the CV mode, the charge current exponentially drops to a predefined termination level where the battery is fully charged and the charging is terminated. Because the LiFePO$_4$ battery has much lower internal resistance, its charging time is much shorter than that for a Li-ion battery.

While a LiFePO$_4$ battery is much safer than a Li-ion battery, a fast-charge safety timer is usually required to prevent charging a dead battery for an excessively long period. The LiFePO$_4$ battery can be overcharged to 4V without safety issues, even though it is specified to charge to 3.6V. However, the energy stored in the battery between 3.6 and 4V is very limited. From the discharge curve in Figure 2.34, the cell voltage drop is very fast at the beginning of the discharge period. This demonstrates that the battery does not store much energy at such a high voltage.

Most of the battery energy is stored near the battery voltage between 2.9 and 3.4V for 1C to 5C discharge rates. Charging the battery higher than 3.6V does not provide much benefit. The voltage difference between the rechargeable voltage threshold and battery charge voltage should be around 200 mV, since it takes a few seconds to drop the battery voltage from 3.6 to 3.5V. Although the LiFePO$_4$ battery has excellent and stable high temperatures, its temperature should still be monitored to ensure safe operation.

The LiFePO$_4$ battery is used in high-discharge current and high-temperature applications such as power tools, electric vehicles, e-bikes, and uninterruptible power supply (UPS) applications. It should not be used for cellular phones and other portable devices due to its energy density limitations.

The LiFePO$_4$ battery charger ICs available include the bq24630 from Texas Instruments and the LTC4156 from Linear Technologies.

2.8 Wireless Charging Technology

Since the release of the Wireless Power Consortium (WPC) specification in late 2010, there have been many implementations of Qi-compliant wireless power technologies that range from the mobile and consumer handheld market to industrial and medical uses. Any electronics powered by a USB port or 5V/19V DC adapters are candidates for Qi-compliant contactless charging.

The market demand for convenient and safe wireless power systems has been growing rapidly. Portable devices such as smart phones, cameras, and game controllers are a few of the battery-powered devices that can be upgraded to a wireless power system to keep their batteries charged without physical contact. While near-field inductive power coupling has been around for some time, applications have been limited to very low power levels. As today's portable devices require more and more power, and as higher power batteries require more advanced control systems, a number of safety risks in the design of wireless power systems have to be considered. Compliance with WPC specifications helps designers avoid safety problems and ensures interoperability among systems.

The basic wireless power system consists of a power transmitter and receiver. The transmitter is usually located in a base station powered by a DC source. The power receiver is located in a battery-powered device that uses the power received to charge a battery. Both the transmitter and receiver contain wire coils, and power is transferred between them without electrical contact through inductive coupling. Since power is inductively transferred from the transmitter to the receiver via coils, the transmitter must power the coil with a switching current that has a sufficiently high frequency to optimize coupling. The receiver coil then picks up the near-field inductive energy and uses rectifiers and voltage-conditioning circuitry to produce DC output. To control the power transfer, it is important for the receiver to constantly communicate with the transmitter to indicate when power is required, how much power to send, and when to stop sending power. These communications data are exchanged through the same coils that couple the power. Figure 2.36 shows the wireless charging block diagram.

The main advantages of wireless charging include the following:

- Lower risk of electrical shock when wet because there are no exposed conductors (for use in, e.g., toothbrushes and shavers).

- Consistent and secure connections: no corrosion when the electronics are all enclosed away from water or oxygen in the atmosphere.

- Convenience: no need to connect a power cable or USB cable; the device can be placed on or close to a charge plate or stand.

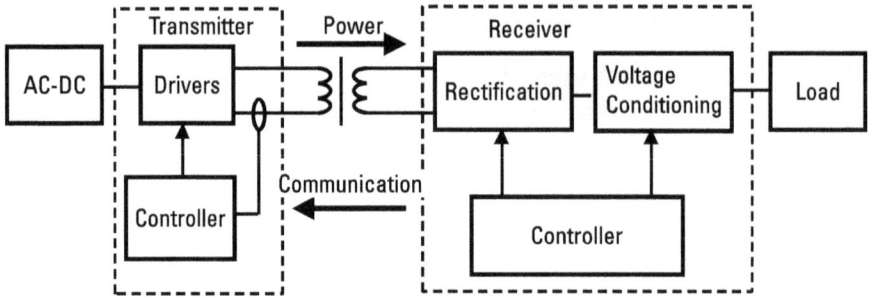

Figure 2.36 Wireless charging system block diagram.

The main disadvantages include lower power conversion efficiency because of the loose coupling between the primary coil and secondary coil. The highest efficiency that wireless charging can achieve is about 72%, much lower than that of wired battery charging efficiency, which is more than 90%. Such a high amount of power loss generates heat within the device. EMI could be another issue due to the loose magnetic coupling between coils. The second drawback is its high cost; it requires drive electronics and coils in both the device and charger, increasing the complexity and cost of manufacturing.

Few wireless charger solutions are available to support the WPC Qi specification. The bq500210 from Texas Instruments is a new Qi-compliant transmitter manager that offers intelligent control of the power transfer between a base station and a mobile device. The bq5101x Qi-compliant integrated receivers provide a regulated DC output and digital-control feedback to the transmitter.

2.9 Solar Charging System

Photovoltaic (PV) systems are low cost, low maintenance, long lasting, and environmentally friendly. They can be constructed to any size in response to the desired energy needs, from a calculator powered by a single cell to off-grid homes powered by multiple full-size panels. Typical PV panels are up to 20% efficient. The energy collected by the PV system varies with the weather. On sunny days, the panels turn out full output power, but on extremely overcast days, the output is small. In other situations, the energy is only needed during the night, for example, for street and home lighting. Most PV systems have batteries. The unused energy produced during the day can be charged and stored in the batteries, making the whole system a reliable source of electric power both day and night, rain or shine. PV systems with battery storage are gradually being used to provide electricity for street lighting and for portable devices such as cellular phones.

A photovoltaic cell, commonly called a solar cell or PV, converts solar energy directly into electrical power. When light strikes the cell, a certain portion of it is absorbed within the semiconductor material. The energy knocks electrons loose, allowing them to flow freely to produce a current. An ideal solar cell can be electrically modeled by a current source in parallel with a diode, as shown in Figure 2.37.

When a load is connected at the solar cell or panel, the actual output current is the current difference between the solar current I_{SC} and the internal diode current. As the load current increases, the current flowing into the internal diode becomes smaller, which results in lower voltage across the diode, such that the output voltage decreases too. In an extreme case, when the load current is equal to the solar cell, where no current flows into the internal diode, then no voltage is generated at the solar terminal.

Figure 2.38 shows a typical I-V characteristic of a solar cell under one light condition and a fixed cell temperature. The solid line represents the output voltage over its output current, whereas the dashed line represents the output power. At open circuit voltage of V_{OC}, the output current is zero, and so is the output power. As the output current increases, output voltage starts to decrease. A maximum power point (MPP) is generated at the specific voltage and current and depends on its light condition and temperature. If the output

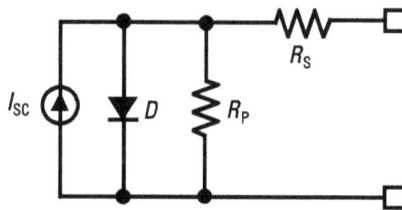

Figure 2.37 Solar cell equivalent circuit.

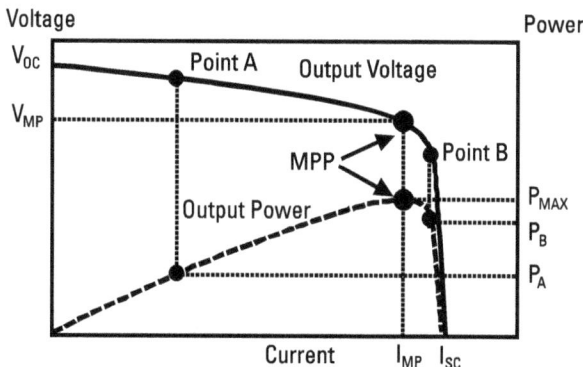

Figure 2.38 Solar cell output I-V characteristics and its output power.

current keeps increasing, the output voltage keeps dropping. As a result, the output power starts to decrease. If the system and the battery charger require a little power, the solar cell may operate at point A as shown in Figure 2.37, where the solar cell delivers power of P_A, less than MPP. This is fine because this is what the system needs. On the other hand, if the charger and system need more current or power, the solar cell may operate at point B in Figure 2.37. The solar cell's output current has increased, but the power P_B delivered by the solar cell is still less than the MPP. A well-designed solar-based charging system should actively control the system to maximize the power from the solar cell or panel.

For most solar panels, when the output voltage drops to around 75% to 80% of V_{OC}, the output power reaches the peak value, which is called the MPP. The MPP operating point is usually provided as a given temperature by the solar panel manufacturer. Because the solar panel's temperature changes significantly under different weather conditions, its MPP operating point is also dependent on the temperature.

The maximum power point (V_{MP}/I_{MP}) specifies the maximum available power from the solar panel. The MPP varies with irradiation, temperature, and other aging effects. A solar charger includes a MPP special circuit that allows the charger to get the solar's maximum power output. A few techniques are available for tracking the MPP by monitoring the solar panel's output current, voltage, and temperature. The instantaneous power can be easily calculated and the MPP can be found through a microprocessor. This method of finding the MPP is very accurate as the temperature and aging effects have been taken into account, but it has a higher cost. The simple and cost-effective MPP solution is based on the fact that the solar cell or panel output voltage is relatively constant, around 75% to 80% of its open circuit voltage at MPP.

This is typically implemented by setting the minimum operating voltage that corresponds with the solar cell's MPP. This MPP charger design allows the charger and system to draw any current from the solar panel as long as the solar panel's voltage remains above V_{MP}. When the output current increases to the point at which the voltage drops to V_{MP}, a special control loop in the charger takes over and regulates the total current from the solar panel to maintain its voltage at V_{MP}. At this operating point, the solar panel delivers its maximum power. As usual, the solar panel provides power to the system first, and any remaining power is used to charge the battery. This voltage-based MPP control algorithm is fairly accurate at providing maximum power, even with varying illumination levels.

Figure 2.39 shows the input voltage-based MPP control block diagram with a synchronous switching buck-based charger. The output of the solar panel is used as input to the MPP solar charger. Besides the traditional CC-CV battery charging control regulation loops, an input voltage regulation loop is

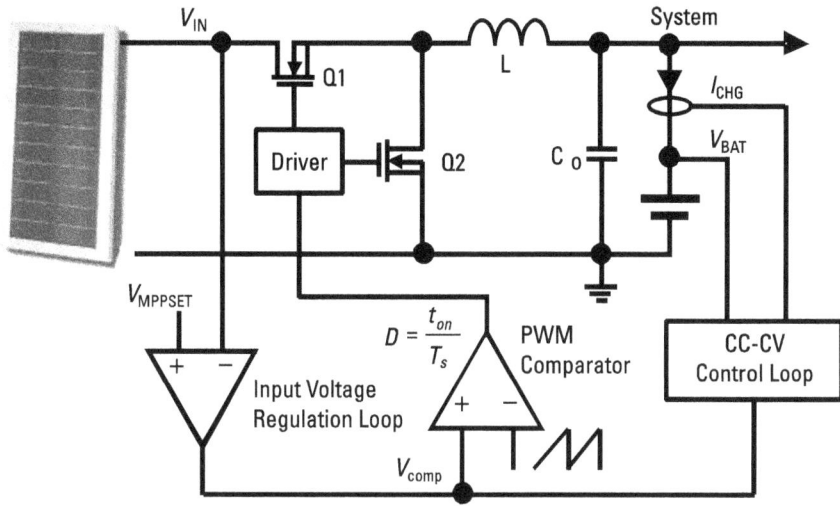

Figure 2.39   Input voltage-based MPP control block diagram.

introduced to make sure that the output of the solar panel is no less than the voltage at the MPP operating point in order to achieve the MPP.

The input voltage-based MPP solar chargers that are available include the LTC3652 from Linear Technologies and the bq24650 from Texas Instruments. The MPP charge controllers integrate the input voltage regulation into the charger integrated circuit for achieving maximum power point tracking (MPPT) besides the CC-CV charging control. The LTC4000-1 has integrated the MPPT charge algorithm with panel temperature compensation, which accurately tracks the MPPT even at different panel temperatures.

References

[1]   Choi, S. S., and S. H. Lim, "Factors That Affect Cycle-Life and Possible Degradation Mechanisms of a Li-Ion Cell Based on LiCoO$_2$," *J. of Power Sources,* Vol. 111, 2002, pp. 130–136.

[2]   Japan Electronics and Information Technology Industries Association and Battery Association of Japan, *A Guide to the Safe Use of Secondary Lithium Ion Batteries in Notebook-Type Personal Computers,* http://www.baj.or.jp/e/news/lithium_ion070828.pdf, April 20, 2007.

3

Battery Safety and Protections

3.1 Introduction

Most people, electrical engineers included, have a deeply entrenched preconception about a battery as an AA primary cell—a metallic cylinder that can be inserted and forgotten. It does not have enough power to do any damage by developing heat, and the only issue most users have experienced is when a battery was left in a device for several years, overdischarged, and leaked. They ended up with a sticky, gooey substance that probably damaged the device in question by corroding everything in it. Because of this intuitively relaxed attitude, an engineer who is going to design modern battery-powered devices must seriously remodel his or her notion of the level of danger a battery represents—or the results can be not only lost jobs, but even lost lives. The purpose of this chapter is to accomplish such a remodeling of battery designers' thinking and to offer a practical solution to all of the challenges that arise from highly energetic mobile power sources.

The requirements placed on modern electronics to become smaller, lighter, and more powerful resulted in battery evolution to a power source that carries a large amount of energy in a small volume. This is true for most commonly used batteries in portable electronics, the Li-ion batteries, and somewhat true for the more rarely used NiMH batteries. Most of the following discussion, however, focuses on Li-ion batteries because they are far more dangerous due to their use of an organic electrolyte. This organic electrolyte is needed because Li (even intercalated) is so active in giving away electrons that it even reacts with water, forcing the electrons into hydrogen and releasing it as a gas, sometimes explosively. So using water is not an option. Organic electrolytes solve this problem because they can only react slowly with Li to form passivating layers on the surface. These layers happen to be (not so much by luck, but rather through the hard work of the chemists who found the right solvents and additives) ionic conductive, so a battery can pass current through them.

However, one of the side effects of organic electrolytes is that when the temperature exceeds a certain limit depending on battery chemistry, the reaction between the active material and electrolyte releases heat and becomes self-supporting (thermal runaway), which can cause a fire, explosion, or venting of electrolyte. Just how destructive can such an event be? It depends on the amount of energy released. Let's compare the energy of an 18650 Li-ion battery with that of a well-known destructive device, say, a hand grenade filled with TNT. TNT releases energy of 4.1 kJ/g. A Li-ion battery stores 0.250 Wh/g, which corresponds to 0.93 kJ. This is about one-quarter of the TNT energy. But—and this is a big difference with traditional batteries that have a water-based electrolyte—a Li-ion battery has additional energy in its organic electrolyte! In case of combustion, it will also burn. So will polymer separators. As reported in a safety analysis [1], the energy released by the burning of up to 10 g of electrolyte and 1.6-g separators contained in an 18650 cell is 280 kJ. Given the cell weight of 40 g, we get 280/40 = 7 kJ/g of combustion energy. So, taken together, the electric and material burning energy per gram of Li-ion battery is around 8 kJ/g, close to twice that of TNT! Comparing an 18650 cell with a typical hand grenade (U.S. M67, with 180 g of explosive), it takes only five 18650 cells to exceed the hand grenade's energy! A typical notebook battery pack (3 serial/2 parallel, 3s2p) has six cells, so it has *more combustion energy than a hand grenade.* I trust this will make you pause a moment.

Fortunately, battery combustion is not nearly as fast as that of explosives; otherwise, in the famous video where a laptop is burning on a meeting room table we would see the table splintered into a thousand pieces all over the room. However, this amount of energy released even within minutes is very likely to ignite surrounding objects, which includes the electronic device itself. A typical notebook computer's plastic casing and printed circuit boards add about 10 times more energy to burning compared to that of the battery itself [1].

I hope I have convinced you that we need to create batteries with all of the best protection circuits, handling practices, and precautions that can be found in the industry, and not to take any shortcuts. For your reference, here are the industry standards that outline such good practices: for multicell packs, IEEE 1625-2008 standard [2] and for single-cell portable applications it is IEEE 1725-2006 [3]. In addition, the Japan Electronics and Information Technology Industries Association (JEITA) guide outlines improved battery charging practices [4]. For some of the failure types that cannot be prevented by pack design, verifying adherence of your cell supplier to cell safety standards becomes most critical. They are covered in the United States by Underwriters Laboratories

(UL) standard 1642 [5] and was scheduled to transition to international standard IEC 62133 in 2012 [6]. In this chapter, we will give you enough detailed background that you will be able to understand the reasons behind these best practices and to utilize them with confidence. We will take a closer look at all of the things that can cause a safety or functionality issue in a battery, so that the strategy to deal with each one of them will become clear.

It is useful to recognize that the triggers for a safety event can be *external* or *internal* to a battery pack, and also *internal to a cell*. If we treat a battery pack as a black box with two terminals (well, actually it is usually at least three, with the third being a thermistor connection for a charger, or five if communication lines are included), external triggers are those that originate from the outside of the black box. For example, the trigger could be too high of a voltage applied by a defective charger, or very high current that happens because the + and − terminals are shorted by a screwdriver forgotten in the same side pocket as the battery. One good thing about external triggers is that we can address them through the design of a battery pack by adding protection electronics that will immediately terminate exposure of our precious cargo—the battery cells—to any mistreatment that is coming from the outside world by turning off the FETs and electrically isolating the battery. Figure 3.1 shows a schematic of a typical notebook (multiple cells in series) battery pack design. We will refer back to this figure as we mention various protection actions.

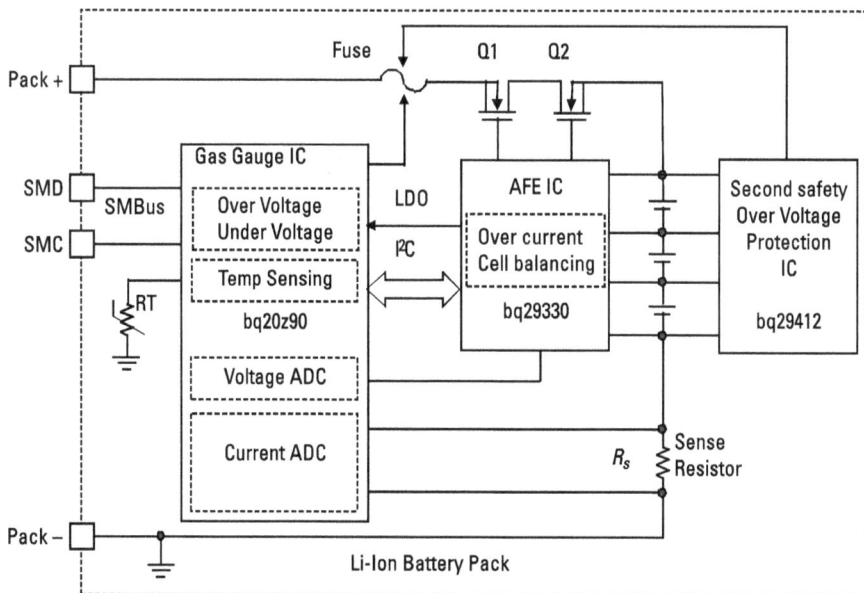

Figure 3.1 Schematic of battery pack protection implemented with a bq20z90 gas-gauge/bq29330 analog front-end (AFE) protector and bq29412 second-level protector chipset.

Pack internal triggers are trickier to deal with because they originate inside the battery pack itself so they cannot be stopped by isolating the external terminals. Such triggers include overvoltage of one cell among multiple cells in series. The charger might be applying the correct voltage to the pack terminals, but the cell voltages are still getting into a dangerous range. These need to be dealt with inside the pack itself, for example, by bypassing some current around the cells. Such management, called *cell balancing*, will be discussed in detail in Chapter 4. Alternatively, charging can be terminated by a protection circuit, again by turning off the charge FET if some of the cell voltages are outside of the safe region even if the pack voltage is correct.

More difficult to deal with are pack internal short circuits, which could happen, for example, is wiring is loose because of a bad solder or spot-welding joint. In addition to strict quality control and design measures to avoid such situations (such as the use of polymer spacers between cells and the coating of all cell surfaces), there are some protections against extreme currents that can be implemented in the cell itself. These types of protection include a positive temperature coefficient (PTC) element that will go high ohmic if the current exceeds the 2C rate, and a high-pressure valve that will break the circuit and insulate the cell if the current is too high. Figure 3.2 shows a schematic of the internal workings of a Li-ion cell.

Finally, the most dangerous and frustrating situation for pack designers is when failures happen inside the cell itself. Sometimes a metallic particle that falls off during some process step, such as during electrode foil cutting, can end up inside the cell. If it gets in the wrong spot where there is no separator or if it cuts through separator, it can create a short between the cathode and anode. This short causes a local release of heat, which activates the reaction between the electrodes and electrolyte, which in turn releases more heat. In some cases this heat release is faster than heat dissipation, and self-propagating thermal

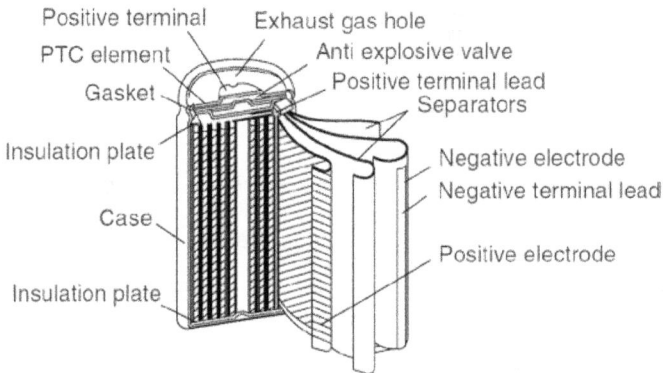

Figure 3.2   Protection features and internal structure of a cylindrical Li-ion cell.

runaway occurs. Nothing we can do in the pack can prevent these types of problems (although some futuristic ideas are being considered). Currently, they have to be prevented by cell makers through strict adherence to industry standards of cell manufacturing, material purity and process control, and use of a cell design that covers and insulates all parts of the metal foil that could cause a short circuit. As a pack maker or system maker, our job will be to find a reliable cell maker, as well as to continuously verify (sometimes audit) that the cell maker does not waver in its commitment to safety.

Now, let's have a closer look at each of these mechanisms by providing a detailed description and discussion of prevention and protection methods.

3.2 Safety Events Triggered External to the Battery Pack

Pack external triggers that can cause an unsafe condition include overvoltage, overdischarge, overcurrent, overtemperature, and a short circuit across the pack terminals.

3.2.1 Overvoltage Applied to a Battery Pack

Overvoltage applied to a battery pack due to abnormal charger operation is the most dangerous condition because it causes a thermal runaway. When pack makers or cell makers want to make a scary video to show how Li-ion cells explode (very popular at battery conferences), they do not have to do anything more complicated than apply to a pack an external charging voltage that results in 5V/cell (for example, 15V for a three-series cell pack) with a decent current level, turn on a camera, and hunker down. In real systems such a situation can be caused by improper system design, use of a wrong charger, or failure by the charger to maintain the correct voltage.

To prevent the problem, system designers have to ensure that the maximal charging voltage is maintained within cell manufacturers' guidelines. It is not always 4.2V per cell (although this number is most common). Please refer to the battery chemistries section for more details on charging voltages.

Connecting the wrong charger can sometimes be prevented by designing a special unique connector (or a connector covered by a standard that provides a fixed voltage, such as a USB). Preventing failure of the charger is a matter of following design standards. Prevention can be made easier if the charger is built around a special charger IC that has been highly optimized and field tested in many designs. These ideas are discussed in detail in the charging section.

Cell overvoltage is just as dangerous as pack overvoltage, and is described in a subsequent section. Table 3.1 summarizes the issues that can cause pack overvoltage and the corrective actions required.

Table 3.1

Pack External Overvoltage

Trigger	Cause	Prevention	Protection
Charger output voltage too high	Improper charger design for given chemistry	Follow cell manufacturer's guidelines	Voltage is measured by protection circuit. If safety threshold is exceeded, charge MOSFET turns OFF, preventing the pack from charging.
	Use of a charger not intended for the battery	Connector design, authentication check	Increased pack temperature can indicate abnormal charging conditions. Protection circuit detects overtemperature, charge MOSFET turns OFF.

3.2.2 Overdischarge

An overdischarge is a situation in which cell voltages go below the manufacturer's specified minimal value. Note that this value is far from zero. For a $LiCoO_2$-based Li-ion battery it is 3V/cell. An overdischarge will not instantly cause a safety event, but it can damage the cells and reduce their lifetime.

It is important to understand that chemical processes are triggered not by the absolute voltage on the cell terminals but by reaching a certain chemical state of charge. During periods of current flow, the voltage at the cell terminals includes both IR drop across the cell and the actual equilibrium voltage that reflects the present state of charge of the cell. During high current spikes, the IR drop could dominate the voltage decrease and cause it to go, for example, all the way to 2V (normally 3V termination is required). A momentary drop to a low voltage will not cause any degradation of the cell because it reflects just temporary nonequilibrium concentrations on the electrolyte side of the double-layer capacitance and is not capable of causing any chemical reaction. Once the current spike is removed, the voltage will recover within seconds to its higher level above 3V. For this reason it would make the most sense to differentiate between "steady-state" undervoltage and "instant" undervoltage. Most protectors are adding some delay to the undervoltage disconnect function to prevent unnecessary system shutdowns during short spikes.

Regardless of load duration, the IR drop creates a difference with the voltage observed external to the cell and the voltage at the surface of the particles where degradation reactions take place. This difference is not straightforward, because active materials are porous and so cell resistance is distributed; it is not just a discrete resistor placed in series with the cell. Basically, there is no single IR drop, but many different IR drops to different positions along the active material thickness. For this reason it is not possible to simply subtract the IR drop completely, but it is possible to subtract the part of it that is not distributed, for example, the high-frequency component that is mostly due to current collectors and electrolyte. A more sophisticated undervoltage function that can react to the actual state of charge of the battery rather than voltage (e.g., eliminate the effect of the IR drop) can be provided by safety systems that are integrated with battery gauging ICs.

It is important to prevent overdischarge not only because it degrades the cells, but also because it could lead to unsafe situations in the future. This is due to dissolution of Cu current collectors that starts at low voltages. Once copper is dissolved during overdischarge, it can be deposited back during subsequent charging. But it will not be deposited neatly as a thin film on the same spots from which it was removed. Instead it tends to grow in the form of dendrites (not unlike tiny stalactites), which can breach the separator, potentially causing microshorts. The effects of microshorts can range from increased the self-discharge rate to a catastrophic thermal runaway. Various reasons for overdischarge and their remediation are summarized in Table 3.2.

3.2.3   Overcurrent During Discharge

An overcurrent situation happens if a device draws current above the manufacturer's recommended rating for various periods of time. Note that the rating depends on the type of cell. Notebook and cell phone-type cells are typically rated for 1C to 2C continuous discharge: 1 hour to 30 minutes until empty (and about twice that for a short pulse discharge). Dedicated power cells, which often use a different cathode chemistry such as $LiFePO_4$ of manganese spinel, can support a 15–20C rate of discharge.

In either case, exceeding the recommended rating will cause overheating of the battery, resulting in problems ranging from minor degradation of cycle

Table 3.2

Pack Overdischarge

Trigger	Cause	Prevention	Protection
Device does not stop discharge at the battery end of discharge voltage	Improper design of the device powered by the battery	Implementing pack voltage-based shutdown of the system	Voltage is measured by a protection circuit. If cell voltage is too low and can damage the battery, discharge MOSFET turns OFF, preventing the pack from discharging.
	Use of a device not intended for the battery	Connector design, authentication check	A fuel-gauge IC can terminate discharging based on SOC information in addition to voltage information to take into account battery impedance, aging, load, and temperature and to prevent terminating discharge too early.
	Device failure: short circuit, power consumption when device is off	Follow design and testing standards	Battery protector terminates the discharge by turning OFF the discharge FET.

life to a catastrophic thermal runaway. Depending on how much the rating is exceeded, the tolerable time decreases. Discharge overcurrent reasons and solutions are summarized in Table 3.3. Typical protection levels and timings used in notebook battery protectors are shown in Figure 3.3.

A special case of overcurrent is a short circuit of the external pack terminals (for example, by a key shorting the input of a pack located in a suitcase). This one is most likely to cause thermal runaway as a result of the extremely

Table 3.3

Pack External Overcurrent

Trigger	Cause	Prevention	Protection
Device draws current exceeding battery maximum specified current	Battery capacity/ power capability insufficient for the device	Consider the maximum power the device draws and battery guidelines for max power. Typical Li-ion cell maximum discharge rate is 2C, unless high-rate cells are used.	*Pack based:* Current is measured by AFE circuit by means of measuring voltage across a sensing resistor. If current exceeds limits for predefined time, discharge MOSFET turns OFF, preventing the pack from discharging . Usually several levels of overcurrent and termination delays are used (see Figure 3.3).
	Use of a battery not intended for the device	Connector design, pack authentication check	*Cell based:* PTC will go high ohmic when current exceeds 2C for nonpower cells. It will recover after cooldown.
	Device failure	Follow design standards	

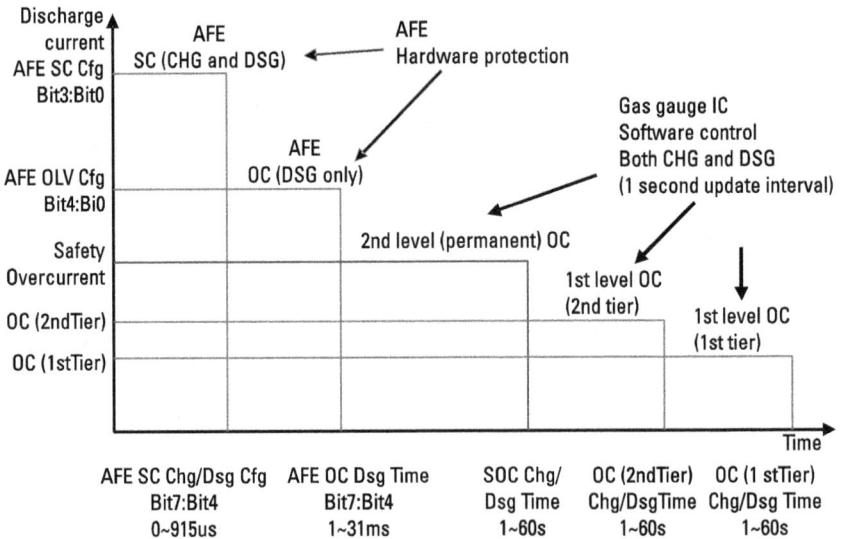

Figure 3.3   Overcurrent levels used in bq20z90 gas-gauge and bq29330 AFE protector chipset.

high currents generated unless protection is used. The current level is usually so high that it requires a microsecond-fast response to avoid damage to the pack's electronics. That is why short-circuit protection is usually handled by different (analog-only) components (AFE IC in Figure 3.1), as opposed to other overcurrent protections that can be handled by more accurate and sophisticated, but much slower digital components (gas-gauge IC).

In come cases ICs can be integrated in the same package or on the same die (silicon slice with components etched/deposited on it). For most IC manufacturing processes, the same die integration is not cost/size effective because digital circuits—processor, memory, ADC/DAC—can be made using a much smaller size process (for example, 90 nm) than analog processes (which typically need 250 nm) because they operate at much lower voltages. If implemented on the same process, analog component requirements would keep all component sizes large, such that overall solutions would be larger than solutions implemented on two different processes for digital and analog. Larger usually also means "costs more" because cost for IC manufacturers is directly proportional to die size. Table 3.4 summarizes the different causes of short circuits and protection approaches.

3.2.4 Overcurrent During Charge

Overcurrent during charge happens if the charger delivers currents above the manufacturer's recommended rating for some period of time. Usually the maximal charge current is recommended to be between 1C and 0.7C rates, although it can be different for cells designed for high-power operation.

Exceeding the recommended limits will cause overheating of the battery as well as lithium plating because a carbon anode cannot intercalate lithium fast enough. Lithium is deposited on the surface of the anode as a metal powder and fiber-like dendrites that are loosely connected with the electrode. Parts of it will gradually get intercalated over time, but other parts will get electrically

Table 3.4
Pack External Short Circuit

Trigger	Cause	Prevention	Protection
Pack terminals are short circuited	Short circuit internal to a device powered by a battery	Short-circuit protection in power supply design	Current is measured by AFE circuit by means of measuring voltage across a sensing resistor. If current exceed limits for predefined time, discharge MOSFET turns OFF, preventing the pack from discharge. For short-circuit detection usually AFEs are used for protection to provide shutdown in the microsecond range (see Figure 3.1).
	Short circuit when battery is outside the device during transportation or handling	Connector design; using a "system present" pin that turns off discharge MOSFET when pack is not in use	

disconnected and will slowly react with electrolyte to produce insoluble products and remove Li from the system, not to mention increase the cell impedance by blocking surfaces and pores. All of these effects are bad for battery cycle life. In addition, Li-dendrite powder is extremely active, so if a microshort happens in the area where the powder exists, it will react with electrolyte at an extremely high rate, resulting in temperatures in the thousands of degrees. For this reason, exceeding the charging rate is added to the list of possibilities that can cause cell-internal thermal runaway and should be prevented independently by the charger and by the system protection circuit.

Interestingly, Li intercalation becomes much slower at low temperatures. For this reason charging below 0°C is disabled for most systems to avoid Li deposition. More advanced charging systems that follow JEITA specifications [4], have different charge rates and charge voltages for different temperatures to prevent Li deposition. The lower rate is typically used below 10°C (Li deposition likely) and above 45°C (degradation accelerates at high temperatures). To support this improved safety measure, the protector should also be capable of stopping the charging process if some of the JEITA charging rates or voltages are exceeded. More details about JEITA charging can be found in Chapter 2. Charge overcurrent reasons and solutions are summarized in Table 3.5. Typical protection levels and timings used in notebook battery protectors were shown earlier in Figure 3.3.

3.3   Safety Events Triggered Inside the Battery Pack

A different group of safety events can be caused without any external influence or with purely mechanical actions, due to failure of different components of a battery pack.

Table 3.5
Pack External Overcurrent During Charge

Trigger	Cause	Prevention	Protection
Charger delivers current exceeding battery maximum specified charging current for given temperature conditions	Improper design of the charger for given chemistry. Use of a charger not intended for the battery	Follow cell manufacturer's guidelines. Connector design, authentication check	Current is measured by protection circuit. If safety threshold for given temperature is exceeded, charge MOSFET turns OFF, preventing the pack from charging. Increased pack temperature can indicate abnormal charging conditions. Protection circuit detects overtemperature, charge MOSFET turns OFF.

3.3.1 Pack Internal Short Circuit

The most dangerous internal failure is a pack internal short circuit. An example of such an event would be the case in which a wire between the cells gets disconnected and then shorts between the positive and negative cell terminals or between the positive terminals and the ground of the circuit. Cylindrical Li-ion batteries have their external casing connected to the negative terminal, whereas prismatic cells have their positive terminal connected to the casing. In both cases, the casing is "active" and it offers a large surface area for attack by the end of a loose wire. Typically, most of the cell casing is covered by insulating plastic film, but there are always some exposed areas. Covering it after assembly, or separating cells physically with plastic barriers that are part of external pack casing, adds more protection. It also helps prevent the cells from moving from their positions, which also could cause a short. Often cells are glued or attached with two-sided sticky tape to the surface of the external pack enclosure to further prevent their movement. Same applies to wires—fixing them to the package surface or even to the cells makes sure that even if they get disconnected, they will stay put and will not cause a short.

Laminated cells (also known as polymer cells or pouch cells, popular in smart phones and tablets) do not have an external metallic casing, but are enclosed in an Al-coated polymer pouch, which acts as insulation. Positive and negative terminals appear as aluminum tabs, which are usually spot-welded to larger nickel tabs suitable for subsequent spot-welding to the printed circuit board of a gauge/protector during pack assembly. This arrangement offers much less surface area for a potential short to the casing. Note, however, that the tabs are usually very soft and can break when subjected to relatively small mechanical stress, and because they are located close to each other, this could lead to a short. Moreover, the polymer pouch is also not very mechanically strong and can be easily punctured by a sharp object. (Some testing videos show how they can be cut with scissors, causing smoke, sparks, and sometimes thermal runaway.) For that reason, laminated cells (single or multiple in series or parallel) need to be protected by a mechanically strong external casing that prevents them from any mechanical stress, bending, or puncturing and also fixes them in place.

Because this type of pack internal short is "outside" of the protection circuit, it cannot be prevented by pack electronics and can cause thermal runaway. Some cell internal protections (PTC and pressure valve) are activated when high current flows through a cell for several seconds (PTC) or minutes (pressure valve). Note that while cylindrical cells are all equipped with PTCs (except for high-power cells because the impedance of PTCs is too high for them), prismatic cells usually need external PTCs. Prismatic cells still have a pressure valve. Finally, pouch cells do not have any cell-level protection unless an external PTC is attached. However, they have some "self-protecting" properties that can be more or less effective depending on the particular pouch cell design. One property is that due to a bottleneck at the edge of the connector tab, very high current can cause the tab to "burn off." In some cases such a bottleneck is created on purpose (for example, in Kokam Co. cells). Another self-protection mechanism is the swelling (we could almost call it "ballooning") that most pouch cells develop if overheating occurs. As the cell is literally turning into a balloon inflated by the gases that develop during overheating, the inflation acts as a pressure-based current disconnect because it creates an empty space between electrodes that stops ionic current. Note, however, that swelling can damage the enclosure, causing shifting of other cells and possibly leading to other short circuits, so it is not generally as clean a protector as a pressure valve. Note also that there is a common misconception that pouch cells do not have a liquid electrolyte and, hence, would not burn. In reality, all pouch cells (even though they are sometimes called polymer cells) do contain the same liquid electrolyte as cylindrical cells although the liquid is sometimes gelated to improve mechanical stability. Gelated electrolyte will, however, liquefy at higher temperature and burn if its ignition temperature is exceeded.

Given the above considerations, a proper mechanical pack design, quality control during manufacturing, and internal cell protections are critical for all types of cells. A summary of pack internal short-circuit causes, prevention, and protections is given in Table 3.6.

3.3.2 Cell Overvoltage

Although it is important to maintain the correct charging voltage at the pack level, the voltage at the cell level is not always equal to the pack voltage divided by the number of cells. Differences can exist initially because the cells' initial state of charge might be different during assembly. This can be prevented by cell grading by voltage and charging/discharging conditioning of the cells. However, even if cells have the same SOC initially, it can gradually change because the self-discharge rate is not the same for all cells. All cells have different leakage rates due to differences in the impurities present and separator microdefects. Additional differences can arise because of the location of the cells in the pack in relation to heat sources. Cells that are closer to the hot spots in the system are exposed to higher temperatures. Because the self-discharge rate is about three times faster if the temperature changes from 25° to 60°C, the hot cells become "lower voltage" cells.

Even if the cells have the same state of charge in the fully discharged state, when charging progresses, the SOC can diverge. That happens if cell capacities are not the same, which is normally the case. Usually pack makers grade the cells according to capacity before pack assembly, so that cells within a pack have less than 1% capacity deviation. However, 1% is still a lot—it means that

Table 3.6

Pack Internal Short Circuit

Trigger	Cause	Prevention	Protection
Short circuit inside a battery pack	Short circuit caused by disconnecting cell leads or wires	Safe practices of highly reliable soldering or spot-welding Additional isolation of cell surface of the cells close to leads Gluing the wires to the external casing or cell casing	Each cell contains a PTC that will shut down in case of overcurrent or overtemperature. Cells contain a mechanical disconnect device (pressure valve) for overtemperature conditions. External PTCs can be used outside the cells. Gauge and protector ICs can detect critical cell imbalances and disable charging to prevent further damage.
	Short circuit on printed circuit board (PCB) of pack electronics	Coat PCB with insulating lacquer Follow layout guidelines	
	Short circuit due to cells shifting out of place	Glue or fix with sticky tape the cells to the external pack enclosure Compartmentalize the cells with insulating separators that are part of the external enclosure	
	Short circuit due to external mechanical stress, a puncture, or other damage	Pouch cells are especially vulnerable. Using a mechanically rigid, resistant-to-puncture external enclosure and affixing the cells to the surface of the enclosure are critical.	

cells that started at the same state of charge will have a 1% capacity difference by the time the cells reach a fully charged state. This difference is enough to create a 20- to 30-mV voltage difference at charge termination, which means that smaller capacity cells are exposed to higher voltages and will degrade even more (since higher voltage means higher degradation). Over multiple cycles this effect will self-amplify as lower capacity cells get lower and lower due to faster degradation, and voltage differences rise higher and higher. Eventually cells can experience such large voltage differences that lithium dendrites form and (in combination with some aggravating event like a microshort) a thermal runaway becomes possible.

The above-described scenario makes it necessary to monitor all cell voltages individually and to take corresponding protective actions to at least prevent unsafe voltage conditions. Basically charging will be terminated by the protector if even just one cell's voltage rises above the safety limits. This is not ideal, because overall pack effective energy is decreased, because all other cells would not yet have reached their full charge when the charge was terminated due to the outlier cell. A better design is one that includes a cell-balancing circuit that ensures that cells have the same state of charge and voltage by the time they reach a fully charged state, even if their capacities are different. That way all cells energies are fully utilized. Such techniques are described in detail in Chapter 4. Note, however, that the cell-balancing circuit is an addition to—not a replacement of—an individual cell voltage-based protector. Typically cell-balancing circuits do not have very large current bypass capabilities because they would be not cost or size effective due to the very small accumulative imbalance that needs to be counteracted. So they achieve balance gradually over time and cannot just instantly enforce it in case, for example, some failure in the cell has caused a sudden change. In addition, because an overvoltage situation can very reliably cause a catastrophic safety event, protection against it has to be redundant: One protective safeguard must take over if the other fails. In fact, in multiple-series packs the industry standards [2] recommend that two independent protectors (gauge/protector IC and independent second-level protector IC) be used, both of which can turn off the charge FET and, if that fails, blow the chemical fuse.

Various scenarios that cause unsafe cell voltages are outlined in Table 3.7.

3.3.3 Cell Internal Short Circuit

Cell internal short circuits have caused the majority of serious battery accidents and are the most difficult to prevent via pack design, because pack protection electronics have no way to influence what happens inside the cell. Unfortunately, there is not much a battery pack or device maker can do to prevent this

Table 3.7
Cell Overvoltage in a Pack with Multiple Serially Connected Cells

Trigger	Cause	Prevention	Protection
Voltage of only one cell in a multiple-series cell pack exceeds safety limits	Capacity differences among the cells	Quality control, sorting the cells by capacity Cell-balancing circuit that keeps the same SOC at the end of charge (even if it diverges at other states)	Each cell voltage is individually monitored. If the limit is exceeded, a first-level protection IC turns off the charging MOSFET to prevent charging. If voltage continues to rise, an independent second-level protection IC turns off the charging MOSFET. If voltage continues to rise, a fuse-blowing circuit is activated by either the primary or secondary protector.
	SOC differences among the cells on assembly	Sorting the cells by voltage Pretreatment that brings cells to the same SOC Improved accuracy of cell formation cycler	
	SOC differences that accumulate during pack operation (different self-discharge rate, microshorts, leakage, cell temperature differences)	Cell-balancing circuit to counteract accumulation of SOC differences Pack and system design that minimizes temperature differences among cells	

type of short, except to select a reputable cell maker and audit the cell maker for continuous adherence to industry standards of cell testing, specifically IEC 62133 [6].

Unfortunately, even the latest versions of cell testing standards focus more on "how badly will the cell fail" after thermal runaway is caused by some well-defined severe abuse such as nail penetration, heating with an open flame, or crashing (e.g., things that can be reproducibly tested) and not on failures caused by internal shorts, which are the cases that result in the most catastrophic failures because no external prevention is possible.

Here, for example, is the list of tests used in IEC 62133 testing:

1. Continuous low-rate charging.
2. Vibration.
3. Molded case stress at high ambient temperature.
4. Temperature cycling.
5. Reasonably foreseeable misuse.
6. Incorrect installation of a cell (nickel systems only).
7. External short circuit.
8. Free fall.
9. Mechanical shock (crash hazard).
10. Thermal abuse.
11. Crushing of cells.
12. Low pressure.
13. Overcharge for nickel systems.
14. Overcharge for lithium systems.
15. Forced discharge.
16. Cell protection against a high charging rate (lithium systems only).

These tests will ensure that when failure does occur, cells will fail in a relatively benign way. Benign does not mean it causes no problems to the consumer, but rather that it causes the fewest number of problems that can be reasonably achieved. For example, a requirement of "no flame, no explosion" exists for most of the abuse tests. Smoke and venting of electrolyte, however, are commonly allowed. So it is important to understand that when a cell passes UL or IES tests that does not mean it is completely safe no matter what. It just means that an industry standard best effort was undertaken to make it as safe as possible, and now design of the protection circuit, external package, and interconnects must ensure that this safety will never be actually challenged because the results of even best effort safety standards are still quite unpleasant and even dangerous to end consumers.

Internal short tests have not been yet standardized because it is very difficult to introduce such a test in a reliable way without breaking the casing and without creating a cell especially for that purpose because such a cell might have different properties from a normal cell. However, work on such test development is ongoing. Meanwhile battery cell manufacturers are working actively on avoiding the presence of any foreign particles in cells and also on isolating all parts of electrodes that might come into contact. Quality of materials, the manufacturing process, and quality control (such as high-voltage tests on the cell stack) are critical to guaranty that a cell will not short circuit.

Some development is ongoing to add "predictive safety" into battery monitoring ICs, which would try to anticipate a cell internal short failure before it happens based on a cell's state of charge and impedance measurements and then disable the charging of the pack long before actual failure. This way a pack will be discharged by the system to bring it to state of "least energy" and then (due to its refusal to be charged) it will be returned to customer service where failure can be analyzed, logged, and the pack safely disposed of. Note that most of the safety tests are done on new cells, whereas aged cells can become less safe. For example, degradation of active material does not happen symmetrically; the anode might be lost earlier than the cathode and this can lead to lithium deposition because there is not enough material to intercalate all of the lithium being extracted from the cathode. Such conditions can be detected by observing differences in OCV dependence on state of charge or by detecting excessively high self-discharge rates. However, when analyzing parameters indicating potentially unsafe conditions, a compromise has to be made in choosing how many "false positives" are acceptable to catch a very small number of "true positives"—packs that would experience an actual thermal runaway. Initially just data collection of such parameters rather than active intervention makes more sense to find an experimentally reasonable threshold settings for such interventions.

Various reasons for cell internal failures and prevention measures are summarized in Table 3.8.

For single-cell battery packs a second-level protector is not usually used, and the gas-gauge and AFE are sometimes combined. Alternatively the gas-gauge is not used and the AFE incorporates over/undervoltage protection along with overcurrent protection. However, the absence of a microcontroller limits the ability to make accurate decisions, which need more safety margins. This results in reduced battery run time. The ability to detect potential safety hazards under a complex combination of factors is also reduced.

Table 3.8
Cell Internal Short Circuits Causing Localized Overheating and Thermal Runaway

Trigger	Cause	Prevention	Protection
Metallic particle	1) Particle is added as an impurity in the active materials 2) Particle separated from the cell's metallic parts during production 3) Particle separated as a result of aging	1) Electromagnetic filters, high-voltage tests prior to addition of electrolyte 2) Manufacturing process control 3) Real-time monitoring during usage, removal of suspect batteries	1) Microporous separator melts down pores above critical temperature, stopping ionic current flow. 2) Pressure release vent removes electrolyte from reaction zone. 3) Ceramic coating on separators reduces short-circuit current and reaction rate at high temperature.
Dendrite of active material	Metallic anode material grows needle-form dendrites	Electrolyte composition Microporous separator Charge rate limits	
Loose wire	Wire leading from electrode to casing disconnects	External insulation of electrode stack	
Direct electrode contact	Electrodes come in direct contact because of damaged separator or mechanical deformation of the electrodes	Robust mechanical design Separators extending on both sides of electrode Separator design that prevents shrinkage	
External damage to the cell	Crashing, penetration, or cutting	Robust casing design	

3.4   Final Thoughts

Now that we have looked in detail at the failure mechanisms that can occur in Li-ion battery packs and corresponding preventive and protective actions, it is useful to distill it into a few useful pieces of advice to achieve safe battery pack design. My favorite ones are as follows:

- Respect the battery. It does not forgive shortcuts.

- Follow standards—this is not optional. Accidents and battery-operated equipment recalls do happen even with experienced OEMs and cell makers and such recalls can cost millions.

- The easiest way to make a battery pack for a portable electronic device with standard protection electronics is to use a reference design from protection IC manufacturers such as Texas Instruments, Maxim, LTC, ON Semi, Seiko, Renesas, and Fairchild. In most cases, all of the protection functions will be supported, although you have to look at the specifics in each case to verify the degree of support.

- Application engineers from these companies are there to help make your design safe and have a lot of experience from which you can benefit.

- Cell maker selection should take into account adherence to safe practices and reputation in addition to satisfying device power, aging rate, and cost requirements.

- If you are not an expert in pack design, hire a company that is, such as Micropower, Microsun, or Nexergy. Such companies will be able to customize a battery pack for your needs. And now that you have finished this chapter, you will be able to evaluate the design with a good understanding of battery safety requirements.

References

[1] Exponent Failure Analysis Associates, Inc.; Mikolajczak, C., et al., *Lithium-Ion Batteries Hazard and Use Assessment Final Report,* http://www.nfpa.org/assets/files/pdf/research/rflithiumionbatterieshazard.pdf, 2011.

[2] *Rechargeable Batteries for Multi-Cell Mobile Computing Devices,* IEEE Standard 1625-2008, http://standards.ieee.org/findstds/standard/1625-2008.html.

[3] *Rechargeable Batteries for Cellular Telephones,* IEEE Standard 1725-2006, http://standards.ieee.org/findstds/standard/1725-2006.html.

[4] Japan Electronics and Information Technology Industries Association and Battery Association of Japan, *A Guide to the Safe Use of Secondary Lithium Ion Batteries in Notebook-Type Personal Computers,* http://www.baj.or.jp/e/news/lithium_ion070828.pdf, April 20, 2007.

[5] Underwriters Laboratories, *Standard for Lithium Batteries, 5th ed.,* UL 1642, 2012.

[6] *Secondary Cells and Batteries Containing Alkaline or Other Non-Acid Electrolytes—Safety Requirements for Portable Sealed Secondary Cells, and for Batteries Made from Them, For Use in Portable Applications,* International Standard IEC 62133, 2010.

4

Cell-Balancing Techniques: Theory and Implementation

4.1 Introduction

In the safety chapter we briefly discussed the issue that when multiple cells are connected in series, the cell voltage is not always equal to the pack voltage divided by the number of cells. How does this happen? This chapter explores that question in detail, but the first question to answer is this: Why do we care?

The first reason is safety. Remember, when lithium ion cell voltage exceeds 4.2V by a few hundred millivolts, it can undergo thermal runaway, melting the battery pack and device it is powering. It can even blow up as a big ball of fire. Although a well-designed pack has an overvoltage protection circuit that will prevent such an event (usually even two independent circuits!), it is better not to tempt fate by triggering this protection unnecessarily.

The second reason is longevity. If the maximal recommended charging voltage is exceeded even a little, it will cause very accelerated degradation. Just increasing the charging voltage from 4.2 to 4.25V causes the degradation rate to increase by 30%. For this reason, the misbehaving cell that has higher than its due share of voltage will degrade faster.

The third reason is incomplete charging of the pack. Let's assume the protector circuit does its job and that charging stops when just one cell gets close to unsafe conditions. Now we have successfully prevented thermal runaway, but all of the other cells now have lower voltages and are not fully charged. If we look at the pack voltage, it will be much less than 4.2V multiplied by the number of cells. Less pack voltage means less pack energy. (It also usually means less available capacity, as we will see later.)

The fourth reason is incomplete use of pack energy. Let's consider another situation. Instead of having too high a voltage, one cell could have too low a voltage compared to others when the pack is close to the end of discharge. A pack protector will prevent overdischarge (which would damage the cell) by stopping the discharge of the whole pack when one cell voltage goes below the cell undervoltage threshold (usually around 2.7V for a $LiCoO_2$-based cell). This means that all other cells are still at higher voltages and have energy left. The pack still has energy, but the device can no longer be used because of one misbehaving cell.

Now that we have established that cell voltage differences are harmful enough to take action to remove them, let's look at the causes of these voltage differences. The first thing to understand is that a voltage difference is not in itself an imbalance, but a manifestation of the differences in state of charge (SOC) of the cells if no current is flowing, and of the cell resistance differences if current is flowing. If we try to instantly eliminate the voltage differences themselves (e.g., the "effect") without eliminating their cause, we will potentially do more harm than good, while wasting hardware resources (cost, size) by overengineering the balancing circuit to provide huge currents that will be required for such an instant result, and wasting energy by unnecessarily passing currents back and forth. Unfortunately, this is exactly what happens in some commonly used balancing schemes that have been designed without an understanding of the underlying mechanisms of the imbalance. The plan is that after reading this chapter you will never design or use such an inefficient system because you will be endowed with a perfectly clear idea of what an imbalance is and how it can be eliminated in the most theoretically efficient way.

4.2 Types of Battery Cell Imbalance That Affect the Charge/Discharge Voltage

4.2.1 State-of-Charge (SOC) Imbalance

A SOC difference is the only cause for cell voltage differences if no current is flowing, known as open circuit voltage (OCV). Indeed, there is a simple correlation between SOC and voltage for any battery chemistry in the form of OCV $= f(SOC, T)$, where SOC is the state of charge and T is temperature. The form of the function is different depending on the chemistries, but in general it is clear that for a given $dSOC$ you get some difference in voltage, $dOCV$. What could be causing these differences in cell SOCs? Let's look at a few possible reasons.

First we consider inaccuracies in the voltage measurement of cell formation cyclers. Most Li-ion cells are cycled after assembly to form a passivating layer on the anode and to detect abnormal cells. At the end of the cycling, all cells should end up in the same state of charge (as indicated by the same voltage). However, cycling equipment is not perfect; there are some channel-to-channel variations that result in cell SOC differences. To reduce these, of course, the cyclers themselves have to be kept well calibrated, but also, after cycling, cells need to go be graded, a process in which the cells are grouped based on close voltage, usually within 2 mV from each other. Now, 2 mV may not sound like much, but keep in mind that cells are stored and delivered in 50% SOC. It happens to be close to the flattest portion of the voltage curve, where a 1-mV difference roughly corresponds to a 1% SOC difference. This same 1% difference in SOC will result in up to a 10-mV difference by the end of charge and a 100- to 500-mV difference by the end of discharge (depending on how deeply the pack is discharged) because the voltage/SOC curve is much steeper in these areas. Note that while the percentage of SOC imbalance remains constant during the entire discharge period, voltage differences among the cells vary with SOC because $dV/dSOC$ varies with SOC. Figure 4.1 shows OCV differences among cells at a constant SOC imbalance but at different states of charge.

This figure shows the dependency for a Li-ion cell. The actual shape of the curve will vary with chemistries, but the concept that SOC differences remain constant regardless of SOC is still valid. Some chemistries, such as a lead-acid chemistry, experience large and almost linear changes of voltage with SOC so it is quite easy to estimate how much $dSOC$ is between the cell for a given dV. Other chemistries, such as $LiFePO_4$, have an almost completely flat voltage profile, so even differences in SOC between the cells as large as 5% to 10% are not noticeable in the cell voltage when it is in the midrange of SOC but can nevertheless cause drastic voltage deviations close to the end of charge and discharge, causing protection electronics to trip and to shut down the charging process before the pack has a change to be fully charged. It is also more difficult to grade such cells by voltage, which makes the need for in-system balance management more critical for this chemistry. But such a system clearly cannot be based just on voltage in this case. Later we will look at systems that work for "flat" chemistries.

The second important reason for SOC differences between the cells in a pack are differences in self-discharge rates between the cells. The self-discharge rate is strongly dependent on temperature. It approximately doubles with every 10°C increase from room temperature. System design does not always take into consideration the need to heat the pack evenly, and places various heat-generating components such as the application processor, backlight, and memory in such a way that they can fit most of the components into the smallest

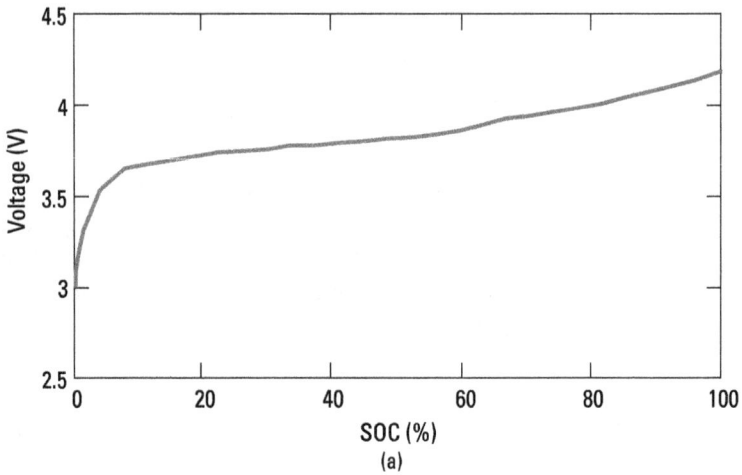

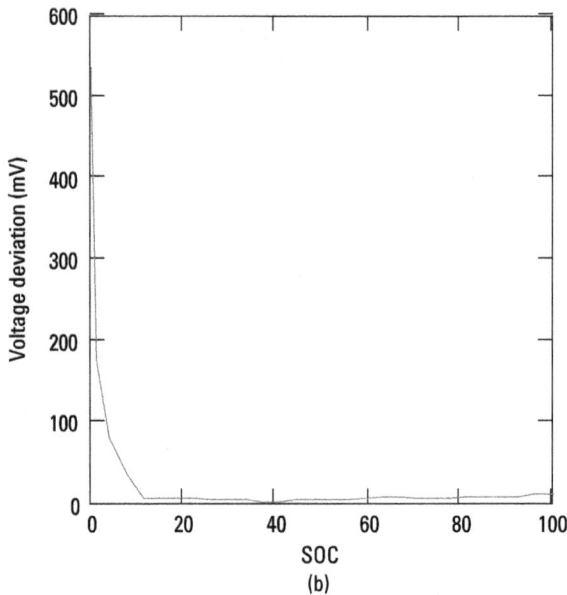

Figure 4.1 (a) OCV dependence on SOC. (b) OCV differences at different states of charge between two cells with a SOC imbalance of 1%.

space, rather than ensuring that cells in a pack will have the same temperatures. This will cause one cell that is hotter to "leak" more charge than a cell that is cooler. This cell's SOC will gradually decrease, because the charger puts the same amount of coulombs into each serially connected cell, but some coulombs in the hot cell get internally short-circuited due to self-discharge and do not contribute to increasing SOC. This process is quite slow, because even at the highest temperature of 60°C, the self-discharge of a Li-ion battery is only about

50% in a year. But over time, the difference in SOC can become significant because the effect is accumulative (the same cell will usually be hotter than the others). Battery pack design could include a heat spreader between the system and the pack, and a good cooling surface on the other side of the cells to keep their temperature and self-discharge in general to the minimum. That also helps to decrease cell degradation, which also happens to be strongly accelerated by temperature.

A SOC imbalance can be also caused by uneven leakage to the battery pack circuit from different cells. It is easier to power pack electronics from the low voltage of just one cell (usually the one closest to the pack ground), because cheaper low-voltage ICs can be used. It looks like a nice shortcut to take, because the electronics would only consume some hundreds of microamps, which is negligent compared to multi-ampere-hour size cells. Unfortunately the effect is accumulative. Over a long period of time, the difference in the amount of charge removed from the lowest cell will increase and eventually reach substantial numbers. For example, if we have three 2,200-mAh cells (Q_{max}), and discharge one by 100 mAh (Q_1), the second by 100 mAh, and the third by 100mAh of actual load + 100 mAh accumulated from 1 month of low current due to powering some circuit, the first and second cells' chemical state of charge will be ($Q_{max} - Q_1)/Q_{max}$ = 95.4%, but the third cell will be at 91%. So we can say that cell 3 is imbalanced by 4.4%. This in turn will result in a different open circuit voltage for cell 3 compared to cells 1 and 2, because the OCV is in direct correlation with the chemical state of charge. This problem can be resolved by good pack design, which powers pack electronics only from the entire pack voltage. It also assures that all of the connections to each serial cell (for example, those used for voltage measuremetns) are high-ohmic. Typically, these connections are ADC inputs with impedance in mega-ohm ranges, and do not draw more than a microamp or two. In addition, this impedance is the same for all cells, so even the tiniest current drawn does not cause an imbalance. All major safety and gauging ICs from Texas Instruments, Intersil, or Maxim are powered from the pack voltage and avoid this issue.

4.2.2 Total Capacity Differences

Sometimes cells have the same voltage at the end of a charge and, hence, appear to be perfectly balanced (and correspondingly have the same SOC), only to show a very large deviation of voltage at the end of discharge. Inversely, cells having an equal SOC and voltage in the discharged state can show large differences at the end of charge. What is the reason for this mystery? It happens to be caused by differences in cell capacities. Indeed, if you have two cups of different heights, they will both be equally empty (e.g., state of charge zero and perfectly balanced) when there is no water in them. But once you pour an equal amount of water into both cups, the taller will be half full, but the shorter will be completely full (e.g., states of charge 50% and 100%, respectively).

The dilemma in such a situation is that you cannot just balance cells with different capacities once and have them stay balanced during charge/discharge (as was the case with SOC balancing for the cells of the same capacity when not given enough time for self-discharge). We have four choices:

1. Balance the cells on top (in the fully charged state) and let them diverge at the bottom. This will be repeatable—they will always stay balanced on the top after that without any further action.

2. Balance the cells on the bottom (in the discharged state) and let them diverge at the top. Again, once the initial balancing is done, it will be a persistent state.

3. Completely rebalance the cells during each discharge, for example, extract energy from higher capacity cells on the way down so that by the time it reaches the bottom all cells have the same energy. This would also require a complete rebalancing during each charge, so that the moment the energy from smaller capacity cell becomes full is delayed until other, larger capacity cells become full. Going back to our analogy with the cups, this would be equivalent to leaking some water from the jar past the shorter cup when poring it so that it will be at the same level as the taller cup at the end.

4. Finally, there is a variation of choice 3, where we are not just throwing away the energy that we take from the larger cell on the way down, but are actually passing it to the lower capacity cells. Inversely, on the way up, energy from smaller capacity cells is not just dissipated, but forced into larger capacity cells. This way, in our example with the cups, we will not end up with two cups at just the 50% level (e.g., matching larger cup), but with both cups at some intermediate higher level, say 75%, because we did not spill any water on the table. It also helps with reducing the subsequent cleaning.

The best choice appears to be obvious (choice 4), but in reality it is not so simple. Because nothing in life is free, passing energy from one cell to another requires an actual hardware implementation (discussed below). Let's just say that it has inductors (or capacitors) and high-power FETs so it is about as large as all of the safety circuits and gauging circuits taken together, so you would likely double the size and cost of your overall battery management solution. Before making a heavy investment in additional hardware, we have to consider how much benefit will we get. That will depend on several factors:

1. How much cell capacity imbalance is there? Little imbalance, little benefit.

2. How large is the cell? Sometimes it is so large and expensive that the additional cost of the balancer would be justifiable to get all of the energy from such a large system.

3. Is heat dissipation (e.g., spilled water on the table) a problem in this particular design because batteries are large and space is confined?

Actually, for many laptop designs it turns out that capacity differences after factory grading can be reduced to 1% to 2%. So in many cases it is cheaper to use a 1% to 2% larger battery rather than investing in doubling the size and cost of hardware.

Now, if we are not going to actually transfer capacity from cell to cell, what is the next best option? Well, closer observation shows that option 3 requires a lot of "spilling of water" both on the way up and on the way down, and it has to happen for every single cycle. Heat dissipation due to damping of additional energy into the environment can be undesirable because it requires installation of additional cooling capacity. But most importantly, what do we gain if we completely rebalance the cells on the way down? All cells will reach 0% at the same time but this will be the same time as the lowest capacity cell would reach it without any balancing. A battery pack protector would have shut down the discharge of the pack at the same time anyway, because the lowest capacity cell will trigger an undervoltage condition. So we released all of this heat from higher capacity cells (at the cost of some additional hardware!) and improved run-time by … zero seconds.

How about the charge direction? There is some benefit to keeping cells from exceeding the recommended charging voltage, so if we can prevent the lower capacity cells from going all the way to the overvoltage threshold (which can be as high as 4.3V for 4.2V normal charging voltage cells), we could improve their health and stop the vicious cycle where the pack gets overcharged, degrades more, ends up with even lower capacity, overcharges more, degrades more …, repeat from the beginning. But do we have to do it every cycle? Actually, if we give up on the idea of rebalancing the cells on the way down (which, as we have seen, has no merit) then we do not have to rebalance them on the way up either! We can just balance it once, on the top, and it will stay this way without any further action! Which brings us to choice 1, which is indeed the most popular choice in portable electronics. Choice 1 is often called *top balancing* and choice 2 is often called *bottom balancing*. To summarize, bottom balancing has no merit in systems that utilize energy dissipative balancing, whereas top balancing helps preventing excessive degradation of a lower capacity cell.

But wait, do you have this question on your mind: If we need to balance only once, why do we even need the balancing circuit in the pack? We could just balance cells on the top to correct the capacity differences during assembly (e.g. "gross balancing") and be done with it, all the while saving money and space. Well, this statement indeed makes sense. We really do not need an expensive and spacious high-current balancer designed to provide such high currents that would allow keeping up with high rate external loads and compensating cell capacity differences in real time. We still need a small lower power balancer to continuously counteract *other causes of SOC imbalance* that are themselves continuous and accumulative in their nature—the differences in self-discharge rate of cells due to their different temperature, for example.

In addition, the cells degrade at a slightly different rate, so cells that had the same capacity initially will gradually develop some "individuality." So this additional discrepancy has to be balanced out, as it develops, in the pack itself. However, both the self-discharge rate and degradation rate are extremely slow and for that reason only balancing current in the milliamp range is needed to keep it in check. Such low current needed for *maintenance balancing* can be easily provided by integrated circuits that are already there, such as protector circuits, without any added cost or size. We look at some specific examples shortly.

4.2.3 Impedance Differences

In addition to SOC differences, another cause of voltage discrepancies among cells is the cell impedance difference. However, this has an effect only if current is flowing. How large is the effect? It depends on cell impedance differences and the current. Internal impedance differences among the cells can be expected in the ~15% range in the same production batch as can be seen form the example test data shown in Figure 4.2(a).

Impedance imbalances do not cause differences in fully relaxed OCV, when any effect from current flow is already dissipated. However, they will

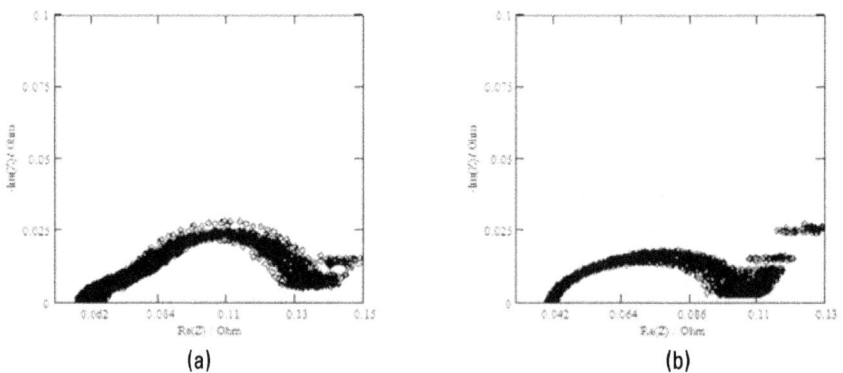

(a)   (b)

Figure 4.2 Impedance spectra differences between 50 cells in one batch for manufacturer (a) and 50 cells for manufacturer (b). Data shown range from 1 kHz (left) to 10 mHz (right).

cause differences in cell voltage during discharge. For steady-state current flow, cell voltage can be approximated as $V = OCV + I * R$, where R is the low-frequency portion of cell internal impedance (right side of the graph in Figure 4.2). If the current is negative (discharge), the voltage will be lower for a cell with higher R. If the current is positive (charge), the voltage will be higher for a cell with higher R.

As can be seen from Figure 4.2, high-frequency impedance at 1 kHz (left side of the graph) is well matched for manufacturer (b) and not as well matched for manufacturer (a). However, low-frequency impedance (the one that will actually matter for continuous discharge) is equally badly matched for both. The reason for this is that most cell makers have access to simple 1-kHz impedance meters that allow them to grade the cell based on high-frequency impedance. High-frequency (1-kHz) measurement is very fast and allows for the detection of massive failures such as a short circuit or current collector disconnect. However, it is not very useful to observe the whole range of electrochemical properties of the battery related to actual charge storage. Low-frequency impedance would be more useful for preventing a voltage imbalance, but its measurement takes at least 10 seconds, so it is very rarely used in cell production (but could be used by a pack maker for improving the cell matching).

No balancing algorithm can eliminate the resistance differences; they are a permanent property of a battery pack once assembled (that is why preassembly grading is beneficial). In fact, the imbalance can increase with aging because different cells' impedance is likely to change at a slightly different rate. But impedance differences need to be considered in any balancing scheme, especially one based on voltage, because they can significantly distort attempts to balance what we can and should balance; namely, the SOC. Note in Figure 4.3 that for the absolute majority of discharges (from 10% to 100% SOC) the distortion of voltage that is caused by the impedance deviation is larger than that caused by a SOC imbalance.

By looking at voltage alone we cannot distinguish which part of the cell deviation is due to a SOC difference and which part is due to an impedance difference. Both parts can shift voltage in the opposite direction! If we do not know about this and just "assume" that all voltage shift is due to a SOC difference, we might be tempted to correct it by bypassing some charge through the cell with the higher voltage that "appears" to have a higher SOC.

However, if most of the difference is caused by an impedance imbalance (as is commonly the case), bypassing more current through this cell will result in the opposite effect—it will increase the SOC difference from other cells to a larger value than it would be without balancing. As a result, the open circuit voltage of this cell at the end of charge will be different from the other cells and can reach high levels, potentially causing the safety circuit to trip.

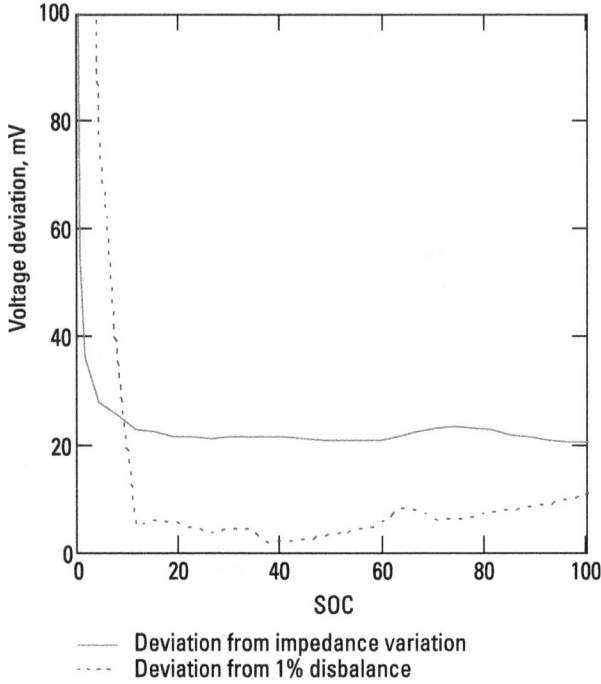

Figure 4.3 *Solid line:* voltage differences between two cells with 15% impedance imbalance at C/2 discharge rates. *Dotted line:* difference between the cells with 1% SOC imbalance for comparison.

If, for example, we are using a simple balancing scheme when a bypass FET allows us to turn on a load parallel to any cell and the bypass FET is turned on based on voltage during charge, it can cause an actual increase of the imbalance through bypassing the cell with the higher impedance. At the end of the charge, the IR rise becomes insignificant because of current decrease, so that the FET switches on at the other cell. However, it happens too late so at the end of charge this procedure results in a higher SOC and higher voltage for low-impedance cells. Eventually it will lead to increased cell degradation. This problem can be reduced if cell balancing only switches on near the end of the charge when the current is reduced and so the I*R drop has a smaller effect on battery voltage.

It is more difficult to fight the IR effect during discharge, because there are usually no predictable periods of low current except for some rare application. Fortunately top balancing (which we earlier found to be the only one useful for "dissipative" implementation) is mostly done only during charging.

There are periods of low current (e.g., in the case of inactivity, rest) that potentially could be used for balancing without IR effect, but it would be still beneficial only if balancing could be done that would ensure the cells were equal on top (at the end of charge), which requires "predictive balancing" rather than "reactive balancing" unless the rest is happening in the fully charged state. See the later discussion on balancing algorithms for more details.

Note that distorting due to impedance differences is inherent to any voltage-controlled balancing method (such as for inductive energy redistribution) and not just to simple bypass balancing. In fact, in energy redistribution cases, the effect can be even more dramatic because energy will be passed back and forth with the high currents that such balancers are typically capable of, which can cause overheating, loss of efficiency, and even complete instability of the control circuit.

Another effect of battery impedance on voltage imbalance exists regardless of any impedance differences. This effect just amplifies voltage differences due to a SOC imbalance. Again, to explain the effect we can model the voltage under steady-state load as $V = OCV(SOC) + I * R(SOC)$ (considering that discharge current is negative). Because function $R(SOC)$ is rapidly increasing its value as SOC approaches zero, the voltage differences between the cells with fixed SOC imbalance increase in highly discharge states, as shown in Figure 4.4. This gives the impression that there is an increased need for balancing near the end of discharge. However, if the SOC imbalance is removed during other stages of discharge and is absent by the time low SOC is reached, the increased

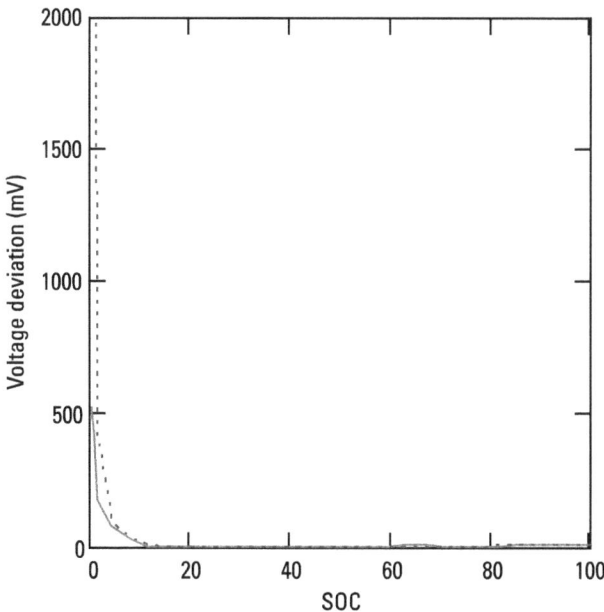

Figure 4.4 Voltage differences under C/2 load at different states of charge among cells with a 1% SOC imbalance. *Solid line:* Differences for OCV case for comparison.

voltage differences near the end of discharge will be eliminated without need for high bypass currents.

4.3   Effect of Imbalancing on Performance

4.3.1   Premature Cell Degradation Through Exposure to Overvoltage

Now that we have a clear understanding of the underlying mechanisms of voltage differences, we can evaluate in more detail different issues that some of these mechanisms can cause. Impedance differences will cause cell voltage differences during charge and discharge, but because this difference in voltage is purely ohmic in origin (to large extent), just like IR drop/rise across a resistor, it does not cause accelerated cell degradation as such and will disappear once the current is turned off. So although there is nothing we can do about this impedance difference, there is also no need to do anything. On the other hand, in the cases of a SOC or total capacity imbalance, the cell with the higher resulting SOC is exposed to higher chemical potentials that will cause accelerated degradation. For example, what happens if one cell has less capacity than the other three serially connected cells in the pack, if they all start at the same state of charge? CC/CV charging will bring the pack to $4.2 \times 4 = 16.8V$ (typical). However, individual cell voltages will not be equal. As you can see in Figure 4.5, the "low-capacity" cell will have a much higher voltage than the remaining cells, while the normal capacity cells will have a lower voltage than is achieved in normal charging. As the cell is exposed to higher potential, it will degrade more, thus

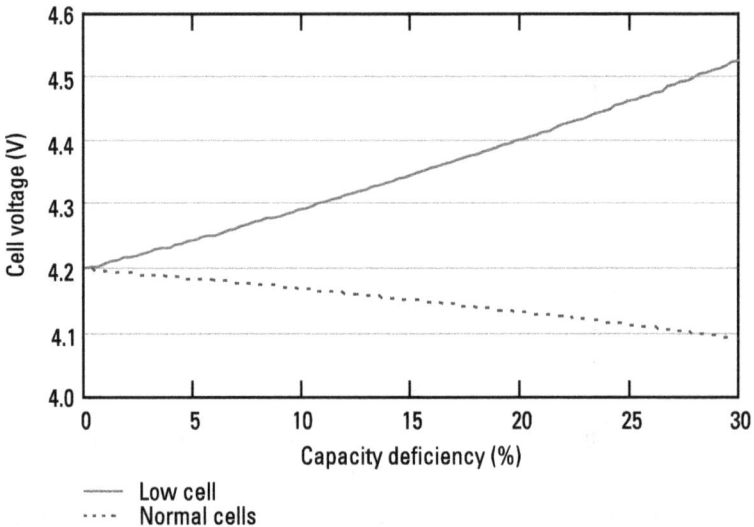

Figure 4.5   Individual cell voltage versus capacity deficiency from nominal.

increasing the capacity deficiency, which will move the pack to the right on the graph in Figure 4.5. Eventually, when the lower cell reaches a total capacity deficiency above 10%, its cell voltage rises into the dangerous area above 4.3V, which will result in extreme degradation of this cell or even become a safety concern.

Note that not all battery chemistries are equally affected by cell voltage imbalances at the end of charge. While the Li-ion chemistry is especially vulnerable because of its ability to store almost 100% of all energy delivered with negligible self-discharge in its operational voltage range, lead-acid, NiMH and NiCd chemistries are relatively tolerant to overcharge because they can respond to increased voltage by internal shuttle reactions that are equivalent to a chemical short circuit inside the cell. For example, in a NiMH battery oxygen and hydrogen generated after the end of charge recombine inside the cell, building water. This causes extensive heating because all the energy of the charger is converted to heat rather than stored, which is undesirable but at least it does not cause thermal runaway. Still, overcharge at high rates does cause increased pressure inside the cell and will accelerate cell degradation and can even create a chance for explosion or venting. The need for cell balancing has to be evaluated in conjunction with rate capability, cooling, and other properties of the charging system.

4.3.2 Safety Hazards Resulting from Overcharged Cells

Li-ion batteries have very high electric energy concentrated in a small volume. While the possibility of its release via a short circuit can be prevented by appropriate mechanical protections, the coexistence of highly reactive chemicals in proximity makes this battery inherently dangerous. Overcharging and overheating of the battery cause the active components to react with electrolyte and with each other, ultimately causing an explosion and fire. Thermal runaway can be caused merely by overcharging a single cell to voltages above 4.35V. Other cells of the pack will also join the explosive chain reaction if one cell is compromised. That is why continuous cell balancing should prevent any cells from getting anywhere near the dangerous voltage territory, and a safety protection circuit should terminate the charge if this somehow happens.

4.3.3 Early Charge Termination Resulting in Reduced Capacity

Additional safeguards present in the battery pack primary protectors (such as bq30z55) and the independent secondary protectors can help alleviate safety issue. The protector and gas-gauging IC will terminate charging if one of the cell's voltages exceeds the programmable cell overvoltage threshold (default 4.35V). An overvoltage protector will terminate charging and prevent an unsafe condition from occurring, but at the same time it will keep the whole pack undercharged (since other cells are now in a much lower state of charge). Overall, the pack loses energy as the result because the higher voltage in the steeply rising area close to a fully charged state does not compensate for the large capacity loss in much more flat areas where the mAh/volt are much higher.

Because the effects of cell degradation caused by imbalance are auto-accelerating, preventing such a vicious cycle allows us to extend battery pack life significantly and to provide a longer run time despite some initial capacity imbalance.

4.3.4   Early Discharge Termination

So far we have mostly discussed cell degradation if overcharged. This is the greatest safety concern and also the most common degradation mechanism. However, the cell will also be damaged if it is severely overdischarged. In fact, if cell voltage goes below 2V, the actual dissolution of the Cu-current collector will occur. This is deadly to the cell if the process is allowed to continue to a significant extent (no current collector, no charge/discharge). For this reason any device using a Li-ion battery absolutely has to have a discharge termination based on voltage. Some naïve attempts to introduce a consumer-replaceable Li-ion battery, for example as a drop-in replacement for coin cells, have so far faltered for the very reason that primary cells that are being replaced did not require any undervoltage shutdown (they just die, being primary) and might keep draining substantial current regardless of battery voltage even when no useful operation is taking place. If a Li-ion cell is placed into a device that was not designed for it, the cell will be overdischarged below 2V and die just like the primary cell that it replaces.

This is mostly not an issue for devices designed to use Li-ion cells, because they will have a full-off state when the pack voltage goes below a safe threshold. In addition, the battery pack itself will often have a battery management system. To prevent overdischarge of cells and resulting damage, battery managements system will terminate the discharge process if any of the cells reach the predetermined low-voltage threshold. The cell-based termination voltage is usually set to a lower value than the pack-based threshold divided by number of series cells, so that the difference can allow for a small imbalance. For Li-ion batteries, the threshold varies from 2.7 to 2.2V depending on the typical discharge rate. However, a larger imbalance will cause the overall pack to terminate when a cell with lowest capacity or SOC will reach cell undervoltage while other cells still have energy left.

Redistributing the energy from higher cells to lower cells during the end of discharge phase can increase a battery's useful discharge time. Note that if "bottom balancing" is being used, it needs to be a redistribution of energy, and not just a bleeding out of extra charge; otherwise, it would just dissipate heat and not increase the run time. To be effective, an inductive or capacitive energy redistribution circuit with high efficiency (usually above 80%) is required. If the control algorithm reacts only to voltage differences that become noticeable only very close to the end of discharge, it would also require a high-rate bypass capability to keep up with the high discharge current. Such circuits are expensive to implement in redistribution balancing circuits, and larger inductors and FETs use up space that is at premium in portable devices. An approach that utilizes the hardware more effectively would be to gradually redistribute any existing SOC imbalance during the entire charge/discharge process, not just when it results in acute voltage differences (at the end of discharge); this is known as *predictive balancing.* This, of course, requires the ability to determine how much charge needs to be bypassed somehow without relying only on voltage. How to accomplish this is discussed in the balancing algorithms section.

4.4 Hardware Implementation of Balancing

4.4.1 Current Bypass

One simple implementation of cell balancing uses a MOSFET in parallel with each cell and controlled by a comparator output for simple voltage-based algorithms that turn on the bypass FETs during the onset of voltage differences, or controlled by a microcontroller for more complex and effective algorithms that can work continuously regardless of variations in the voltage. A general setup is shown in Figure 4.6.

The main choice here is to use MOSFETs that are integrated in the balancing controller IC and typically have bypass currents from 9 to 2mA (depending

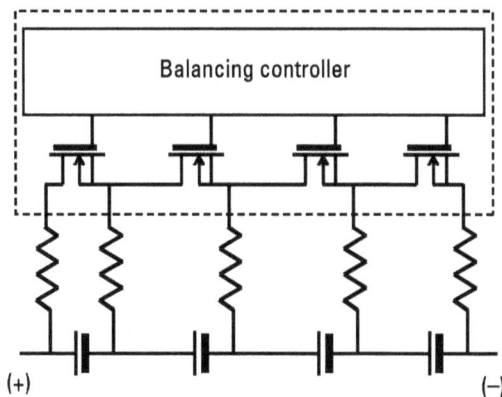

Figure 4.6   Cell-balancing setup using bypass FETs.

on the choice of the external resistors), or to use external FETs with bypass capability that can be freely tailored to particular application needs.

In Li-ion batteries that have a very low self-discharge capability and, therefore, an accumulative imbalance per cycle of usually less than 0.1%, the bypass current of internal FETs is sufficient to keep the pack continuously balanced. In other chemistries where self-discharge rates are much higher and, therefore, differences in the self-discharge rates among the cells result in higher SOC differences per cycle, higher rates might be needed. Some balancing circuits have separate pins for voltage measurement and charge bypass; however, this is not common in portable devices because the larger number of pins increases the size of the device, which is a disadvantage in constrained spaces. The issue with balancing current interfering with voltage measurement is usually addressed by the firmware turning balancing FETs on at a time when measurements are already finished.

Passive cell balancing using integrated FETs is limited by low balancing current and, therefore, may require multiple cycles to correct a typical imbalance. To achieve fast passive cell balancing, an external bypass circuit can be implemented by modifying the existing hardware. Figure 4.7 shows a typical implementation. The internal balancing P-MOSFET S_N for a particular cell, which needs to be balanced, is turned on first. This creates a low-level bias current through the external resistor dividers, R1 and R2, which connect the cell terminals to the battery cell balance controller IC. The gate-to-source voltage is thus established across R2, and the external MOSFET S_{EN} is turned on. The on-resistance of the external MOSFET S_{EN} is negligible compared with the external cell balance resistance R_{BAL}, and the external balancing current, I_{BAL}, is given by $I_{BAL} = V_{CELL}/R_{BAL}$.

By properly selecting the R_{BAL} resistance value, we can get the desirable cell-balancing current, which could be much higher than the internal cell-balancing current and can speed up the cell-balancing process.

The drawback of this method is that balancing cannot be achieved on adjacent cells at the same time, as shown in Figure 4.8. When internal MOSFETs S_N and S_{N+1} of the adjacent cells are turned on, there is no net current flowing

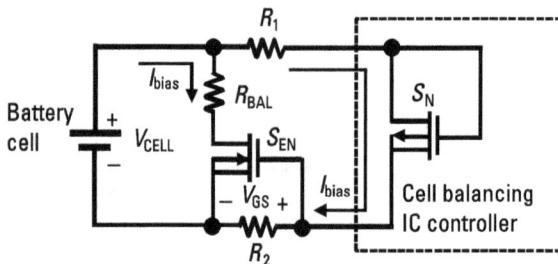

Figure 4.7   External passive cell-balancing circuit.

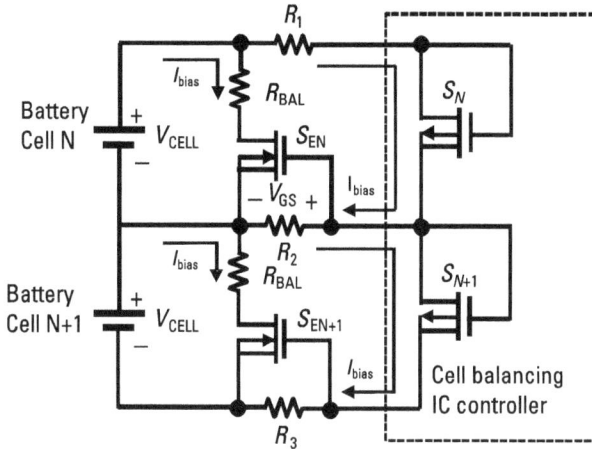

Figure 4.8 The adjacent cell-balancing issue.

through R2 and no voltage drop is created across R2. So there is no gate-source bias voltage of MOSFET S_{EN}, which remains off even when the internal MOSFET S_N is on. In practice, this is not an issue because the fast external cell balancing can quickly balance the cell associated with S_{EN2}, and then the cell associated with S_{EN1} will be balanced. Therefore, the adjacent cells cannot be balanced at the same time, and only every other series cell can be balanced with this approach at the same time. However, some cell balancing ICs have two pins per cell—one for voltage sensing and one for balancing—which makes for a simpler circuit that can balance all cells simultaneously, but such ICs handle fewer cells or have a larger number of pins.

4.4.2 Charge Redistribution

The disadvantage of the current bypass approach is that the energy of the bypassed charge is wasted. While this can be acceptable during charge while the system is connected to a power grid, during actual usage of the battery in portable applications every milliwatt-hour is precious. This makes desirable a cell-balancing approach that would allow us to drain the "high" cells to the bottom in the most efficient way.

The ultimate approach to accomplish this is to use a pack that has no serially connected cells at all. The step-up converter then ensures that the device obtains sufficient voltage. This way, energy waste as a result of the cell-balancing process is completely eliminated. The trade-off, however, is lower efficiency of the power supply, as well as increased size and complexity.

Other solutions can include circuits that allow to for the transfer of energy from high cells to low cells rather than burning it in a bypass resistor. Note that use of the correct control scheme is still critical even in this case because all circuits have limited efficiency and if a charge is unnecessarily shuttled back and forth multiple times due to, for example, an IR effect on voltage, overall efficiency could go close to zero after multiple "swings" even if the single-pass redistribution efficiency can be as high as 80%. So all of the balancing algorithms discussion in subsequent chapters apply to charge redistribution circuits just as they apply to bleed balancing.

4.4.3 Charge Shuttles

One simple approach for redistributing the energy among cells is to connect a capacitor first to the higher voltage cell, then to the lower voltage cell, as shown in Figure 4.9(a). More complicated implementations allow us to connect not only two nearby cells, but also several series cells, as shown in Figure 4.9(b).

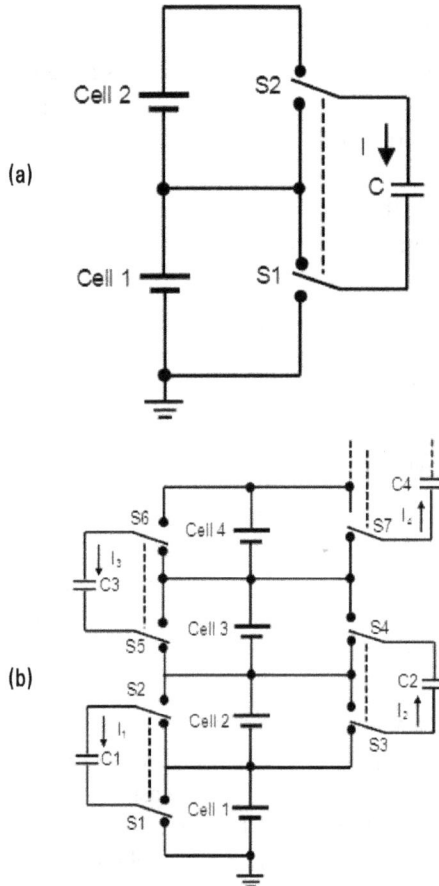

Figure 4.9   (a) Simple capacitor-based shuttle cell balancing circuit. (b) Charge shuttle circuit with several series cells.

Cell 1, cell 2, ..., cell n share flying capacitors with their two neighboring cells, so charge can travel from one end of the cell string to the other. This approach would take a large amount of time to transfer charge from the high cells to the low cells if they are on the opposite ends of the pack because the charge would have to travel through every cell with time and efficiency penalties. This would not be an efficient solution.

Energy loss during capacitor charging is 50%, so heating in the FETs used as switches has to be considered if high-current balancing is supported. However, because there is no charge loss with this process, the energy available on the pack terminals decreases only due to the decrease of cell voltages. Another problem is that high voltage differences between the imbalanced cells exist only in highly discharged state. Because this method transfer rate is proportional to voltage differences, it only becomes efficient near the end of discharge or the end of charge so the total amount of imbalance, that can be removed during one cycle, is low.

4.4.4 Inductive Converter–Based Cell Balancing

Active cell balancing overcomes the energy loss of the passive approach by using capacitive or inductive charge storage and shuttling energy to deliver it where it is needed most. This can be done with minimum energy loss if combined with an optimal balancing control algorithm that allows us to take full advantage of a circuit's inherent redistribution efficiency by avoiding back-and-forth shuttling. It is preferable for efficiency-conscious designs and for applications where delivering the maximum run time is top priority.

A switch-mode power converter concept can apply to the cell balance for achieving the energy transfer from one energy source to another. Figure 4.10 shows the active cell-balancing circuit based on the switch-mode power conversion concept.

A MOSFET, a diode, and a power inductor are composed of a buck-boost converter to complete a charge transfer between an adjacent pair of cells as shown in Figure 4.10. This is a bidirectional buck-boost converter, which can transfer cell energy from either direction. Figure 4.11 shows the switching waveforms of the inductor current and cell-balancing current. If the cell-balancing control algorithm determines that the top cell N needs to transfer its energy to the lower cell, the SN signal, operating at a few hundred kilohertz with a certain amount of duty cycle triggers to turn on the P-MOSFET S_N. The voltage of top cell V_{CELLN} applies to the inductor L_N and the inductor current linearly increases. The cell energy is first transferred from the top cell to the inductor during this time period. When the SN signal resets, S_N is turned off at t_1, and the energy stored in the inductor reaches a maximum value. Because the inductor current must flow continuously, the diode D_{N-1} is forward biased

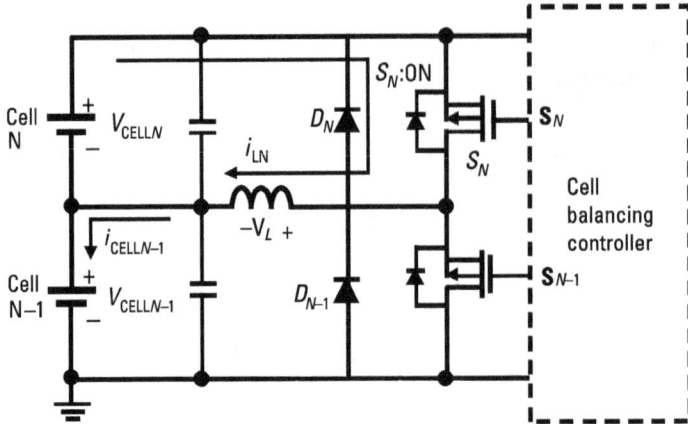

Figure 4.10 Buck-boost converter based cell-balancing circuit implemented in a bq76P-L537A active cell-balancing IC.

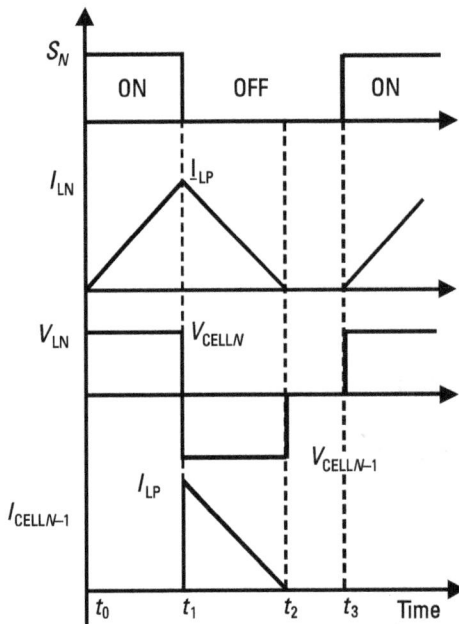

Figure 4.11 Switching waveform of active cell-balancing circuit.

and a negative cell voltage $V_{CELLN-1}$ is applied to the inductor, which results in an inductor current decrease and transfer of the energy stored in the inductor to the lower cell. When the inductor current reaches to zero at t_2, all energy stored in the inductor has been completely transferred to the lower cell, and the diode is naturally turned off with the minimum loss. If the cell $N-1$ has more energy than that of the top one and needs to transfer its energy to cell N, switch S_{N-1} is turned on first and the energy from cell $N-1$ is stored in inductor L. When switch S_{N-1} is turned off, then the energy stored in the inductor is transferred to the top cell N through the diode D_N. In this energy transfer process, the energy loss includes loss from the series resistance of the inductor, and the diode, and switching loss of the MOSFET.

Overall, 90% power transfer efficiency can be achieved with such active cell balancing. The balancing current is determined by the inductance switching period and its turn-on duty cycle. The current level could be much higher than passive cell balancing and more efficient. Besides the obvious advantages, the beauty of such cell-balancing technology is that balancing is achievable regardless of the individual cell voltages.

Figure 4.12 shows the active cell-balancing circuit with N series cells. From this circuit, it is found that the energy can only be transferred from the top cell to the lower adjacent cell or from the lower cell to higher cell as well.

As we know the flyback converter is the isolated power converter of the buck-boost converter. Its output is isolated from the input, and output can be floating such that it can connect anyplace. Figure 4.13 shows an active cell-balancing circuit that can transfer the energy from the bottom cell to the top cell directly with a flyback converter. When switcher S1 is turned on, the bottom cell voltage is applied to the primary winding. The current flowing through the magnetizing inductor linearly increases and its energy is stored in the magnetic field. When switcher S1 is turned off, the energy stored in the magnetizing inductor is released to charge the top cell through the output diode D_n. Therefore, the extra energy from the bottom cell can be transferred to the top cell through a flyback converter. The main limitation of such an active cell-balancing method is that energy is only distributed to the adjacent cell, not to any target cell. On the other hand, such cell-to-cell balancing is only good for battery packs with few cells in series. For long strings, due to the inefficiency of the converters at each step of the transfer, too little energy can be transferred from one end of the pack to the other, making active balancing less efficient than dissipative (passive) balancing.

How do we charge the weak cell or discharge the strong cell to achieve cell balancing? After measuring the voltage or capacity for all cells, the average voltage or capacity can be calculated. The cell with the biggest deviation from the average could be identified. The cell with the lowest voltage or capacity can be recharged from the pack, while the cell with the highest voltage or capacity could be discharged. Figure 4.14 shows the synchronous switching bidirectional flyback converter used to achieve active cell balancing for any target cell. Assume cell 2 is recognized as the weakest cell, and it needs to be charged. Switcher S_p is turned on first, the whole pack voltage is applied to the primary winding, and it stores the energy in the magnetizing inductor of the

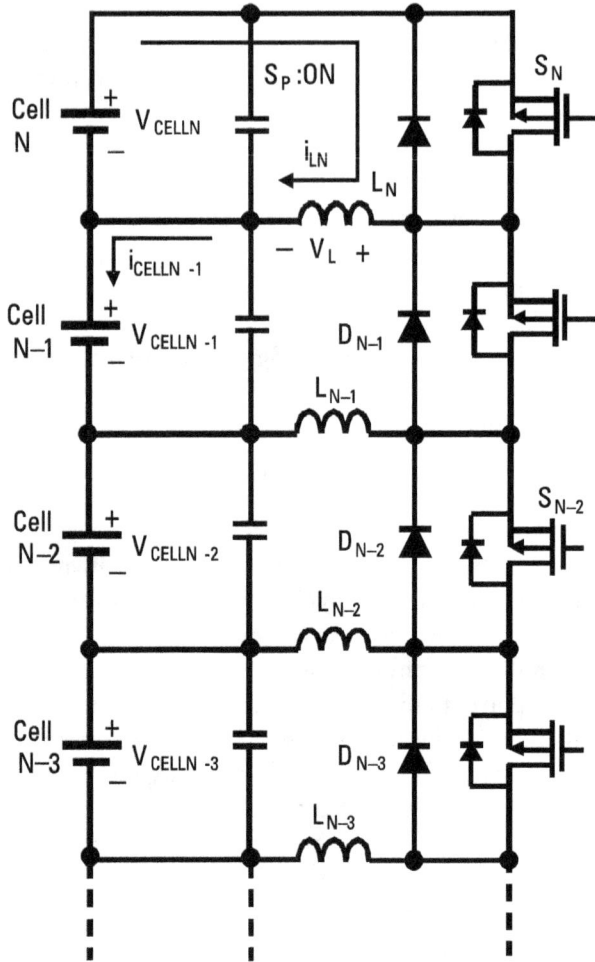

Figure 4.12   N-series cell-balancing circuit.

transformer. When switcher S_p is turned off, the MOSFET S_2 associated with the selected weakest cell is turned on. The stored energy of the transformer can be transferred to cell 2. This can extend the battery pack's run time by preventing the weakest cell from reaching the end of discharge earlier than that of the rest of cells. Therefore, this operation can achieve the cell balancing required for transferring energy from the pack to any either of the cells.

On the other hand, if is more effective to shift energy from a strong cell to the pack. Without cell balancing, the charge process has to stop immediately when one cell reaches it maximum cell voltage, even though the other cells have not been fully charged. Assume that cell $N-1$ has been detected as the strongest cell in the battery pack, and we need to discharge the energy from cell

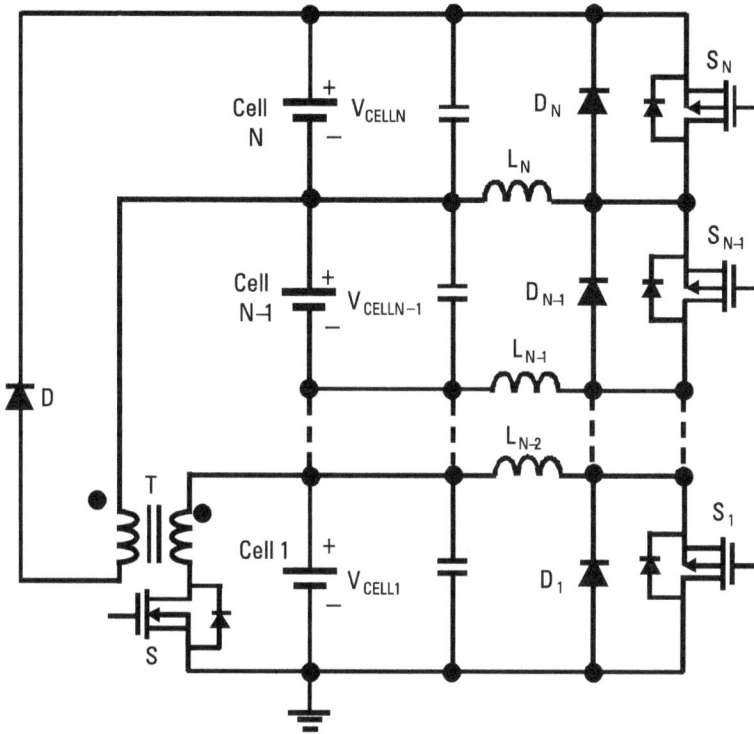

Figure 4.13 Flyback-based cell-balancing circuit for transferring the energy from the bottom cell to the top cell.

$N-1$ and redistribute it into the pack. MOSFET S_{N-1} is turned on first. The cell voltage is applied to the winding connected to cell $N-1$ and stores the energy from the cell $N-1$ in the magnetic field. Once MOSFET S_{N-1} turns off, MOSFET S_p, connected in the primary winding, is turned on so that the energy stored in the magnetic field is transferred back into the pack through the primary winding. So, this operation can achieve cell balancing by transferring cell energy to the pack.

4.5 Balancing Algorithms

Regardless of the particular hardware implementation, there is always a decision to be made regarding when to turn on a bypass switch or when to engage the energy exchange circuit for a particular cell. Different algorithms used to make this decision are reviewed in this section. For simplicity we will refer to the case of current bypass because application of the logic to other balancing schemes is trivial.

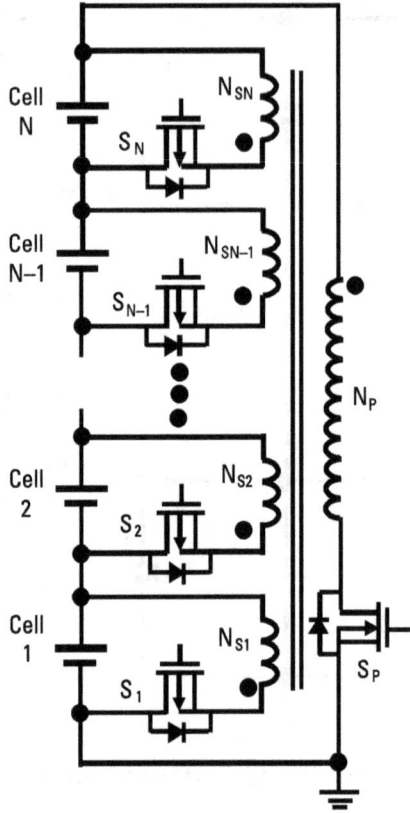

Figure 4.14 Synchronous switching bidirectional flyback converter-based active cell-balancing circuit.

4.5.1 Cell Voltage Based

The simplest algorithm is based on the voltage difference among the cells. If that difference exceeds a predefined threshold, bypass is engaged. To resolve several problems that accompany this simple method, more complicated modifications can be implemented if a microcontroller is used to execute the algorithm:

- *Balancing during charge only* is used to save energy in portable applications.

- *Balancing at high states of charge only* is used to decrease the effect on SOC balancing that can come from an impedance imbalance, because current decreases during the CV mode until charging terminates on minimal taper current.

- A variation of above, in which the balancing is enabled only at a low current regardless of the SOC.

• *Simultaneous multicell balancing* makes decisions about which cells have to be bypassed in terms of the entire pack, not just neighboring cells, as is the case with comparator-based solutions.

One advanced implementation of voltage-based algorithms that incorporates all of the above optimizations is used in the bq2084 battery fuel gauge. Figure 4.15 shows the voltage convergence of multiple cells during balancing.

4.5.2   SOC Based

If a method for determining SOC that is independent of the voltage being under load is available, the balancing algorithm can be improved, because it is no longer vulnerable to impedance variations. However, if no independent method to measure each cell's full capacity exists, equal capacity has to be assumed for all cells such that the capacity imbalance will not be considered. Such a method works as follows:

1. Determine the initial SOC for each series cell bank separately. One of the determination methods is to use an open circuit voltage correlation with the state of charge. This method can only be implemented in a microcontroller with flash memory and significant computational resources because of the need to evaluate voltage versus SOC function OCV(SOC,T) in real time.

2. Determine how much charge is needed for each cell to reach a fully charged state. This requires knowledge of total capacity, which is assumed equal for all cells.

3. Find the cell that has the largest amount of charge needed to reach full capacity, and find the differences dQ among all other cells that need a charge and that of the largest one.

4. This difference has to be bypassed for each "excessive" cell during one or multiple cycles. To achieve that, the bypass FETs are turned on during charging for the duration of each cell's calculated bypass time. The bypass time is calculated dependent on the value of the bypass current, which in turn depends on values of bypass resistance, R_bypass, because time = dQ * R_bypass/(V_average * duty_cycle). Although not a very accurate estimate of needed bypass time, this method is acceptable for low-rate bypass, because during many cycles of balancing, the amount of needed balancing time will be recalculated after every cycle.

5. Alternatively, the bypass current around each cell can be continuously integrated I = V_{cell}/R_bypass) and bypass FETs are turned off once

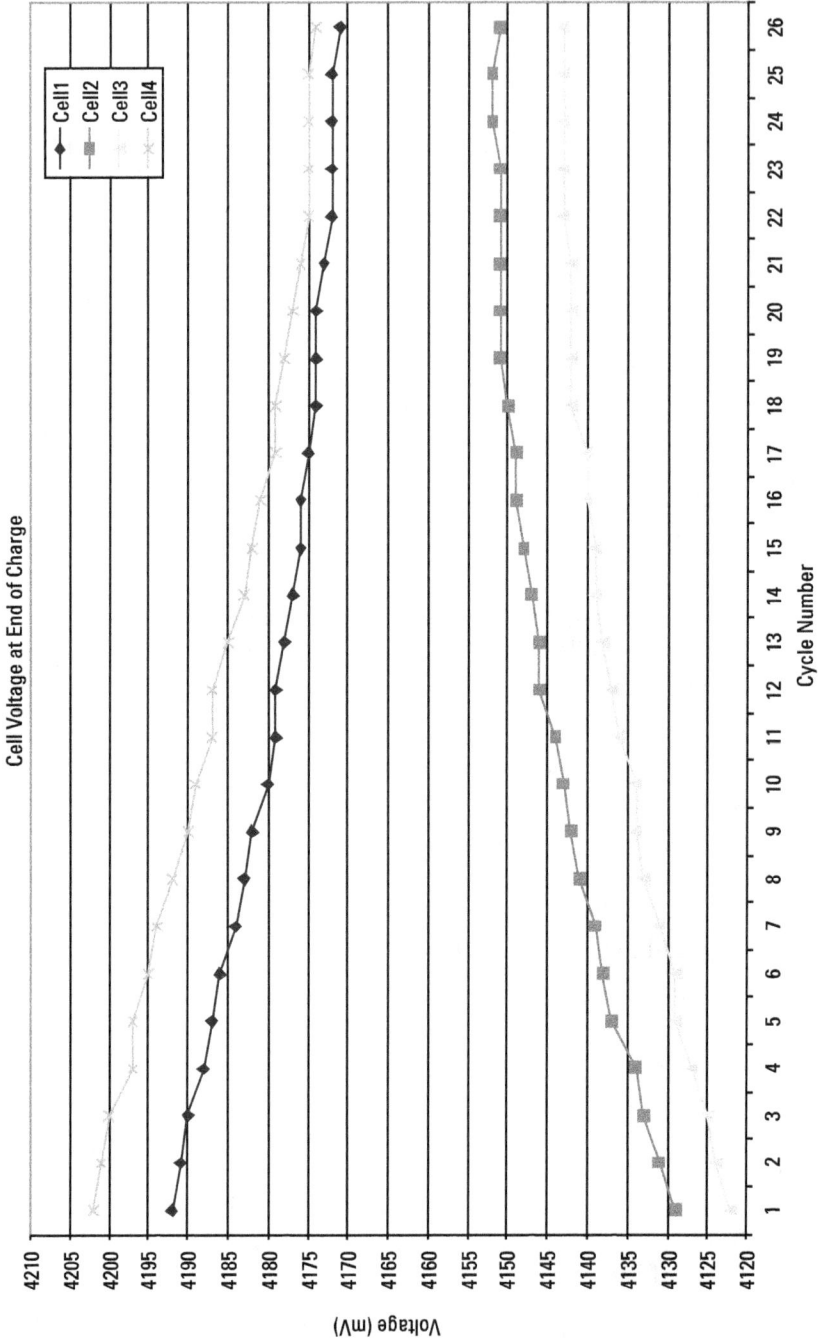

Figure 4.15   Cell open circuit voltages of a four-cell pack at the end of charge during balancing. Initial imbalance = 10%.

the needed bypass charge has passed. This method is preferable if high bypass currents, capable of balancing cells in one cycle, are used.

4.5.3 SOC and Total Capacity Based

Further improvement of the above method takes into account differences in total cell capacities (Q_{max}). This becomes possible if the cell-balancing algorithm is an integral part of the more complex gas-gauging algorithm that is monitoring the state of each cell and capable of measuring changes in the total cell capacity of each cell, as is the case, for example, with bq20z80 cell balancing. Overall, the balancing method is similar to that described in Section 4.5.2, except for calculation of dQ, which now takes into account each cell's individual capacity. Figure 4.16 shows the progress of the cell-balancing process, which causes changes in the open circuit voltage at the end of charge in a three-cell pack, where cell 1 was individually discharged and cell 2 was charged by 2% prior to the test. The bypass resistance used in this test was 700 Ω.

4.6 Summary

In this chapter, we analyzed three main mechanisms that can cause voltage differences among cells that are serially connected in a pack: (1) a SOC imbalance, (2) a total capacity imbalance, and (3) an impedance imbalance. Because the ability to add or remove only a certain amount of charge is available to

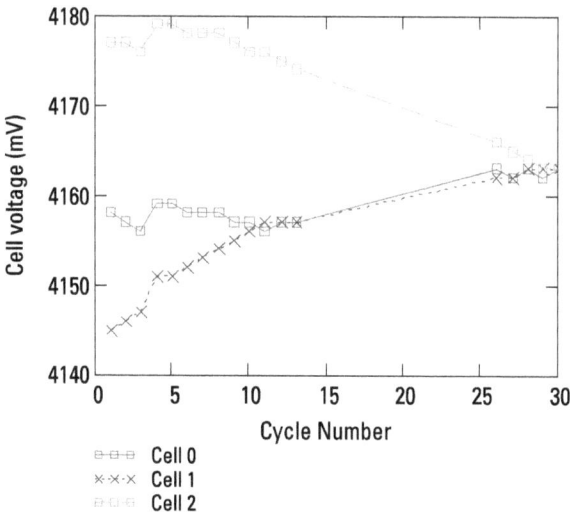

Figure 4.16 Evolution of cell voltages during SOC/Q_{max} balancing, starting from an initial 2% down (cell 1) and 2% up (cell 2) imbalance.

balancing algorithms, only the first type of imbalance, an SOC imbalance, can be eliminated. The second type of imbalance (capacity) has to be taken into account in bypass charge calculations and the third (impedance) should be kept in mind as a distortion, if voltage is used as the balancing criteria, to improve the balancing process and prevent the introduction of a larger imbalance.

Considering the low self-discharge rate of Li-ion cells, we can conclude that if continuous balancing is engaged, the use of integrated FETs provides sufficient balancing current. Use of external FETs may be required if voltage-based balancing is used; that is only active in the areas where a SOC imbalance is reflected by high voltage differences (mostly at the end of discharge). Active balancing methods can provide higher efficiency, but are not at present cost effective for portable applications. An exception might be the case in which extremely high reliability and longevity of the battery pack are needed, because active balancing extends the usable life of a pack, primarily due to the complete use of pack energy regardless of the amount of imbalance.

Voltage-based balancing algorithms have the advantage of simplicity of hardware and implementation, but suffer from slower balancing rates and the possible introduction of additional imbalance through distortions from impedance differences. SOC and total capacity based methods are more complicated to realize but can take advantage of the already present gas-gauging capability of controller ICs and ensure, for given bypass capabilities of the hardware, the most accurate and fastest balancing possible.

5

Battery Fuel Gauging: State of Charge, Remaining Capacity, and State of Health Indication

5.1　Introduction

We called this chapter "battery fuel gauging" because it is the expression that everyone is using to describe "what this little chip inside the battery is doing." However, as we will show in the following pages, it is doing a lot more than that. The origins of the expression are easy to understand—just as we have a fixed amount of gasoline in a gas tank, we have a certain amount of energy in a battery. Early automobiles had actual barrels attached to them, so it was easy to look at the barrel, see the level of the gasoline, and estimate about how far you could go. In the modern digital world, however, more definite information is needed so it is customary to report everything exactly to several decimal points and with accurate units. So why don't we report the actual number of gallons of gasoline left in a car instead of sticking with the representation of an arrow ominously approaching a thick red line on the dashboard?

The reason is that in the case of car fuel, the intuition that a certain amount of fuel equals a certain number of miles or kilometers traveled is oversimplified. Clearly, different cars have different efficiencies. A Toyota Prius can travel 50 miles on 1 gallon of gas, whereas grandpa's old Chevy truck (or a super-race car with a 700-hp engine) will only go 10 miles on the same amount of gas. So to get to the actual useful quantity "miles" we need to divide available energy by the "energy consumption per mile." Fortunately, such calculations are not

139

needed while driving, because traditionally we put larger tanks on cars with less efficiency. For example, the Prius has only a 10-gallon tank, whereas large trucks have 20-gallon tanks. What is kept (relatively) constant between the cars is the total number of miles they can travel. For this reason seeing that you have a little space left between the arrow and the thick red line on the dashboard fuel display means approximately the same travel time for different cars, which is a more useful piece of information than number of gallons.

However, we must consider an additional complication. The efficiency itself is not constant. If car is going up a hill, it will use more gas per mile than in downhill driving. A heavily loaded car will also use more gas, as will a car driven in stop-and-go traffic. For this reason the "miles remaining" estimate has an inherent uncertainty. Just showing the fuel level is a fair indication of "something" and it relieves a car-maker of the responsibility to be accountable for the actual correctness of a statement such as "You have one mile left" that is made to a driver who is stuck on the side of the road one mile before a gas station!

Initially, the makers of electronic devices did not bother with any capacity indication at all. The idea was that everybody has many cheap AA batteries laying around and a new one can be plugged in at any time when the old one is dead. That was fine for devices such as a cassette player or a remote control because you did not lose anything when such a device unexpectedly shut down. The situation nowadays, however, has changed in two different ways.

First, truly portable devices with high energy consumption have appeared. Sony Inc. was the first to commercialize a portable video recorder that required a tremendous amount of energy (compared with what AA cells can offer). The option of carrying around suitcases full of AA cells was considered briefly but rejected. Fortunately, at about the same time the Li-ion intercalation cathode and anode were developed by Prof. Goodenough and Samar Basu, respectively. These were combined with an appropriate electrolyte and separator by Akira Yoshino and perfected to the point where an actual manufacturable Li-ion battery cell was produced by Asahi Corporation. Sony Inc. realized that this was exactly the missing piece of the puzzle and in collaboration with Asahi organized mass production of Li-ion rechargeable batteries that had the energy needed to power a camcorder while being small enough to fit inside it. It had only one problem—it was so expensive that you could only afford one of battery because its cost was comparable with that of the entire camcorder!

With the high cost of the battery, the situation with electronic devices became more similar to that with cars—if you ran out of fuel in the middle of recording an important event (for example, a wedding), you would be in a situation similar to that of a motorist stuck on the road side near a dark forest with raging wolves all around. So a need for capacity indication suddenly became critical. Sony Inc. successfully answered the challenge with a somewhat accurate gauge that was equivalent to the car dashboard fuel indicator. Instead of using fuel level, however, they looked at the voltage of the battery. However, battery voltage itself is not linearly related to energy. To follow the analogy with the liquids, the voltage level in a Li-ion battery would correspond to a liquid level in a strangely formed wine glass of exactly the form shown in Figure 5.1.

As you can see in Figure 5.1, if you start filling this glass with "charge" liquid, it will almost immediately fill the "leg" until the 3.5V level is reached, but then the level growth will slow, coming almost to a standstill at around the 3.75V level. After a substantial addition of liquid, the level will start growing quickly again after 3.95V.

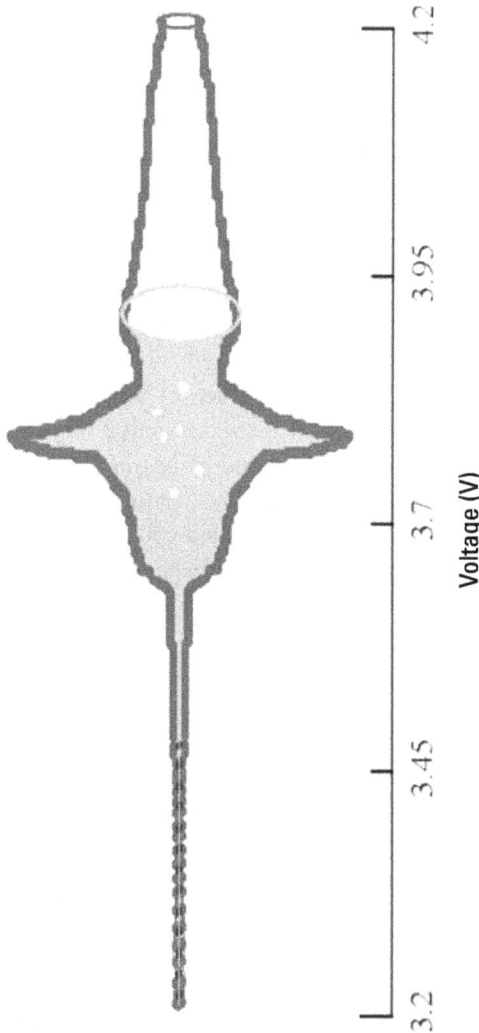

Figure 5.1 Shape of a vessel in which the level of a liquid would correspond to voltage in a Li-ion battery, where the volume of the liquid would be the battery charge.

Sony Inc. solved the problem of converting the voltage into state of charge by using a microcontroller-based gauge, because it would be too complex to resolve the correlation between voltage and capacity of this shape by means of a discrete circuit.

The trouble, however, only started with the camcorder. As the computing revolution was gaining steam, computers began to shrink from the whole-building size to whole-room size to "table-top" size and finally ended up at a size that could conceivably be carried around. That was not quite true, however, for computer power sources—the lead-acid car battery just would not fit. An attempt to solve the problem with nickel-metal hydride batteries appeared to succeed—they fit into a box that could fit into a briefcase. But it turned out to be so heavy that the first users felt as if the briefcase was full of bricks. Again the Li-ion battery came to the rescue and the portable computing industry was born. However, along with the compact size it brought the same problem as we experienced with the camcorder—the battery was so *expensive* that *only one was affordable.* The issue of running out of power in the middle of an important meeting while giving a presentation (e.g., once again alone in the forest with the wolves) had again appeared.

To make matters worse, portable computers were being used as data-processing devices, leading to another issue that was even more severe than the first—the issue of loss of information. Imagine working on an article or school homework or, in my case, on a book (which by a remarkable coincidence just happens to be in the middle of an intercontinental flight). You just finished typing the most interesting idea and suddenly the screen turns deadly black and your hard work is gone forever. Not only could the things that were done in the past hour be lost, but even earlier data could be corrupted because hard drives and flash memories do not like to experience unexpected shutdowns. In addition, data-processing electronics consist of a fragile collection of nanoscopic structures in silicon that do not like sudden current spikes or drops; shutdowns need to occur under controlled conditions, otherwise permanent damage is possible.

To address the issue of data loss and electronics protection, the idea of a soft shutdown appeared. What if we do not allow the system to go to the breakdown point when it dies uncontrollably as voltage becomes too low, but instead develop a controlled shutdown procedure and initiate it when we have just as much energy left as needed for the shutdown procedure itself? In this case we avoid data loss and electronics damage, while shortening the usable run time just as little as possible. That was a noble idea but it run into an unexpected obstacle—the battery capacity indicators need to be able to alert the system to the moment when only shutdown energy is left. Since early capacity indicators showed only the relative "level" of capacity compared to the full capacity, it was not clear how this level corresponded to actual energy needed.

Effectively the portable computer makers were back to the issue that prevented the car makers from reporting "remaining miles"—except this time it had to be solved exactly. There was no excuse for inaccuracy because this time the situation was equivalent to shutting down the car based on information on its gauging indicator that 1 mile is left. Do it too early, and some of your battery is wasted. Batteries are very expensive, so waste is not appreciated by the purchasing department of your company and even less by a user who is trying to finish his presentation. Do it too late and data loss and possibly even device damage occur. To realize how dramatic this issue is, we need to remember that this drastic action would occur quite commonly because running out of battery power is quite a likely event (happens on every flight), as opposed to running out of gas in a car, which usually happens just once in a lifetime (because your spouse is going to kill you if it happens twice).

All of the preceding challenges made it necessary to develop an indicator that could provide an accounting of the exact amount of remaining energy and that would be accurate under any variable load and temperature conditions. This is a remarkably complicated problem and this chapter explores its solution in detail.

The need for highly accurate battery gas gauges is now widespread because most of the devices we use these days share the same issues as laptops: They require an *expensive battery*, are used for *data processing*, and need to have their *fragile electronic circuits protected*. However, the way to achieve high accuracy is not obvious. Superficially, accurate gas gauging might sound like a hardware requirement—precision of voltage and current measurements and resolution of used ADCs. In this sense modern battery monitoring ICs have achieved remarkable improvements. Tiny battery monitor ICs from Texas Instruments as well as a few others have achieved accuracies as high as ± 1 mV for voltage and 0.1% for current measurements. These accuracies exceed many research-grade table-top battery testers. However, it turns out that additional complications, such as battery-operated devices with highly variable loads or peculiarities of battery voltage response, require sophisticated analysis of collected data so that accurate predictions of usable capacity can be made. That means that significant on-board processing capability and new generations of science-intensive firmware are decisive for achieving a reliable battery monitor. This chapter provides an overview of traditional battery gas-gauging methods and discusses the principles of new generation self-adaptive methods.

5.2 State of Charge and Accuracy Definitions

Use of SOC indications that affect device operation requires a definition of SOC that is most relevant for the particular device's purposes. Such a definition is not as obvious as it appears. Different vendors of fuel-gauging ICs might define SOC differently and so device operation has to be adjusted accordingly or an IC with suitable definition needs to be selected.

The simplest definition is based on the maximum possible discharge capacity that can be achieved at a low rate of discharge, Q_{max}. To find Q_{max} (see Figure 5.2) the battery is charged to full, then discharged at a low rate (such as C/20), until the minimum voltage defined by the battery manufacturer (typically 3V/cell) or minimal system voltage is reached, while integrating passed charge Q_{passed}. The value of Q_{max} is equal to the integrated charge at the point at which the termination voltage is reached. State of charge is then defined as

$$SOC = \left(Q_{max} - Q_{passed}\right)100/Q_{max} \qquad (5.1)$$

A "full chemical capacity" based definition is simple, but it is suitable only for very low discharge rate applications where the internal resistance of a battery does not noticeably change the cell voltage. The situation is different for high rate applications, as illustrated by Figure 5.3. You can see that the $I * R$ drop is causing battery voltage to be much lower, so the termination voltage is reached earlier. This means that capacity available until the termination point is reached will be less than Q_{max}. This actual application capacity is called *usable capacity*, $Q_{use} < Q_{max}$. Note that the usable capacity is called *full charge capacity* (FCC) in the smart battery specifications (SBS) common in the notebook world, so it will be referred to as such sometimes. The choice of reporting the usable versus chemical capacity also affects SOC estimation, because to be useful as an indicator of a device's remaining operational ability, at the termination voltage

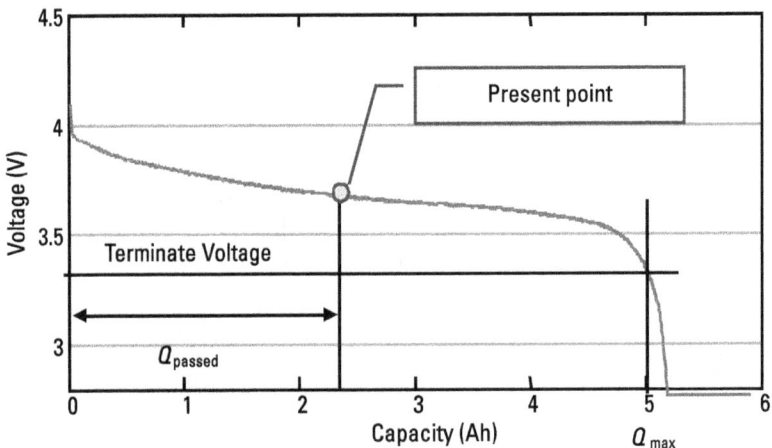

Figure 5.2 SOC definition of low rate of discharge.

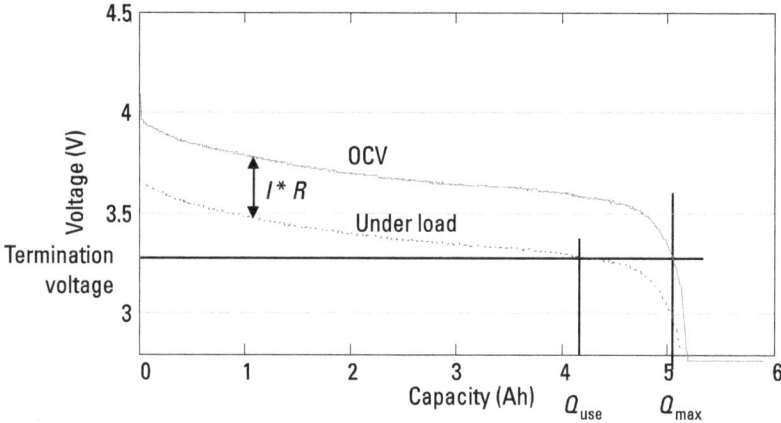

Figure 5.3 SOC definition for an arbitrary rate of discharge.

point SOC needs to be zero regardless of the rate of discharge. So the SOC definition changes from that of (5.1) to one that is relative to Q_{use}:

$$SOC = \left(Q_{use} - Q_{passed}\right) * 100 / Q_{use} \qquad (5.2)$$

The same equation needs to be used for the case where the temperature changes significantly. At low temperatures R is much higher, so higher IR drop across the battery internal resistance will reduce Q_{use} compared to Q_{max}.

We mentioned above the importance of accuracy. But what is accuracy and how is it defined? When selecting a fuel-gauging solution, different accuracy claims can be found in the documentation. Each claim has to be considered together with its error measurement definition, but note that measurement accuracy does not equal gauging accuracy (although used sometimes as such in documentation). For example, accuracy of current integration itself does not necessarily reflect the error in SOC because other error contributions such as the rate and temperature effects are present.

The simplest and most adequate error measurement is to compare the SOC reported by the fuel gauge with true SOC. The difference between the two will be the error. True SOC is defined as follows:

$$SOC_{true} = \left(FCC_{true} - Q_{true}\right) / FCC_{true} \qquad (5.3)$$

Here Q_{true} is the integrated charge by a calibrated reference device, and FCC_{true} is the integrated charge at the moment when the termination voltage is reached at a given rate. Note that the termination voltage is the voltage at which a particular device will shut down. Accuracy depends on the choice of the termination voltage (a lower termination voltage will result in a lower error because $dSOC/dV$ is higher at lower voltages). For this reason it is important to perform an accuracy test with the actual termination voltage you are going to be using in your device. For the same reason it makes sense to take any accuracy claims that do not indicate battery chemistry or termination voltage used with some suspicion.

Because FCC_{true} only becomes available at the end of discharge, the error analysis can be only done after the discharge process has finished, based on the log data. Voltage can be plotted versus the true and reported SOCs to indicate the error graphically, as shown for example in Figure 5.4(a). The error itself is given in Figure 5.4(b).

An important issue is the accuracy dependence on parameter updates, and the conditions under which updates will (or will not) happen in case of a particular device usage pattern. Such relationships will be discussed for each particular fuel-gauging method. Most older fuel-gauging methods depended on parameter updating at the end of discharge (which will not happen at all in many devices because of hibernation to save data) or close to the end of discharge (which will happen very rarely). An error increase in the absence of parameter updates is an important characteristic of a capacity gauging method that needs to be kept in mind.

SOC can also be defined in terms of energy. Indeed most electronic devices require constant power for their operation, so as battery voltage decreases during discharge, current will increase and the same amount of charge will result in less run time at the end of discharge. So remaining power is a better indication of run-time estimation than remaining charge. SOC can be defined in terms of energy similar to the way it is defined in terms of charge:

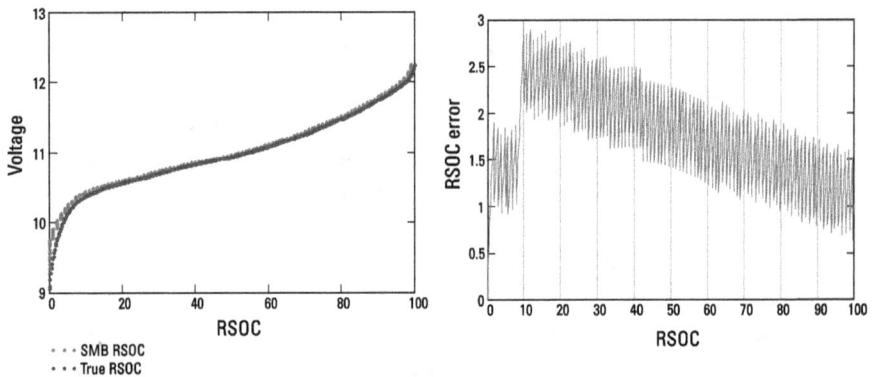

Figure 5.4 (a) Discharge voltage plotted versus reported and true SOC values. (b) Difference between the true and reported SOC values.

$$SOC_e = (E_{use} - E)*100/E_{use} \qquad (5.4)$$

Here E_{use} is usable energy measured by integrating the $V*I$ along the discharge curve at given discharge power load until the termination voltage is reached, and the E charge is integrated from the beginning of discharge until the present moment. SBS for notebook fuel gauging allows reporting of both energy and capacity. However, many gas gauges do not report a truly integrated energy and instead simply multiply the milliamp-hours by average voltage such as 3.6V/cell, which does not reflect changes of voltage at increased current due to higher IR drop, which causes changes of energy at different rates of discharge. This will introduce a significant error in the reporting of remaining energy.

5.3 Basic Battery Remaining Capacity Monitoring Methods

5.3.1 Voltage Correlation

Voltage-based methods were one of the earliest gas gauge approaches because they require only voltage measurement across battery terminals. These methods are based on the known correlation between battery voltage and remaining capacity as shown in Figure 5.5.

It seems to be straightforward, but the battery voltage correlates in a simple way with capacity only if no or very light load is applied during measurement. When a load is applied as it is in most cases when a user is interested in the capacity, battery voltage is distorted by the voltage drop due to the inter-

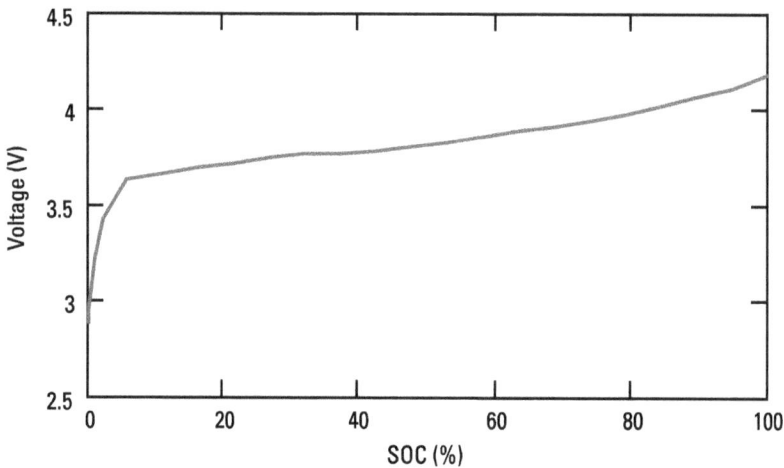

Figure 5.5 Correlation between battery voltage and state of charge for LiCoO$_2$/graphite-based Li-ion battery.

nal impedance of the battery. Methods that apply resistance corrections can somewhat improve accuracy and are discussed in the following sections.

5.3.2 Voltage Correlation with IR Correction

Correction of the voltage drop based on the knowledge of battery impedance using the formula $V = \text{OCV} + I * R(\text{SOC}, T)$, where OCV is battery open circuit voltage, is possible given the impedance dependency on state of charge and temperature. Some naïve attempts at compensation using fixed resistance cause drastic errors. Let's assume that we are going to correct voltage under load by subtracting $I * R$ drop from it, and then use corrected voltage to obtain the current SOC. The first problem that we will encounter is that R depends on SOC. If we use an average value, it will introduce an error in SOC estimation of up to 7% for 1/2C rate, as can be seen in Figure 5.6.

A solution for this would be to use a multidimensional table of voltages at different loads depending on the SOC. The resistance also strongly depends on temperature. It increases about 1.5 times with every 10°C temperature decrease, as can be seen in Figure 5.7. If this dependence is also added to the table, the use of the table becomes quite computationally expensive as well as larger. Gathering the data for an accurate table such as this is also demanding.

An additional complication appears because battery voltage does not immediately react to change of load, but has a delay. The reason for the voltage response of batteries is in its complex electrochemistry. The charge has to travel through multiple layers of electrochemically active material storing the energy (anode or cathode) first in the form of electrons until the surface of the particle is reached and then in the form of ions in the electrolyte. These chemical steps can be associated with time constants in battery voltage response that range from milliseconds to thousands of seconds. It means that after applying a load, the voltage will gradually decrease with time at a varying rate, and gradually exhibit a recovery after the load is removed. Figure 5.8 shows such voltage relaxation after applying a load to a Li-ion battery at different states of charge.

Considering the transient behavior of battery voltage response means that the effective R will depend on time of load application. Otherwise, treating internal impedance as simple ohmic resistance without considering time will lead to significant errors even if we consider R(SOC) dependence from a table. Because the slope of the SOC(V) function depends on SOC, the error is higher in flat portions of the voltage and lower in steep portions. Correspondingly, a transient error will range from 0.5% in the end of discharge state to 11% around 30% SOC (Figure 5.9).

Additional complication is impedance variation among different cells. Even newly made cells are known to have low-frequency (DC) impedance variation ±15%. That makes a significant difference in voltage correction at

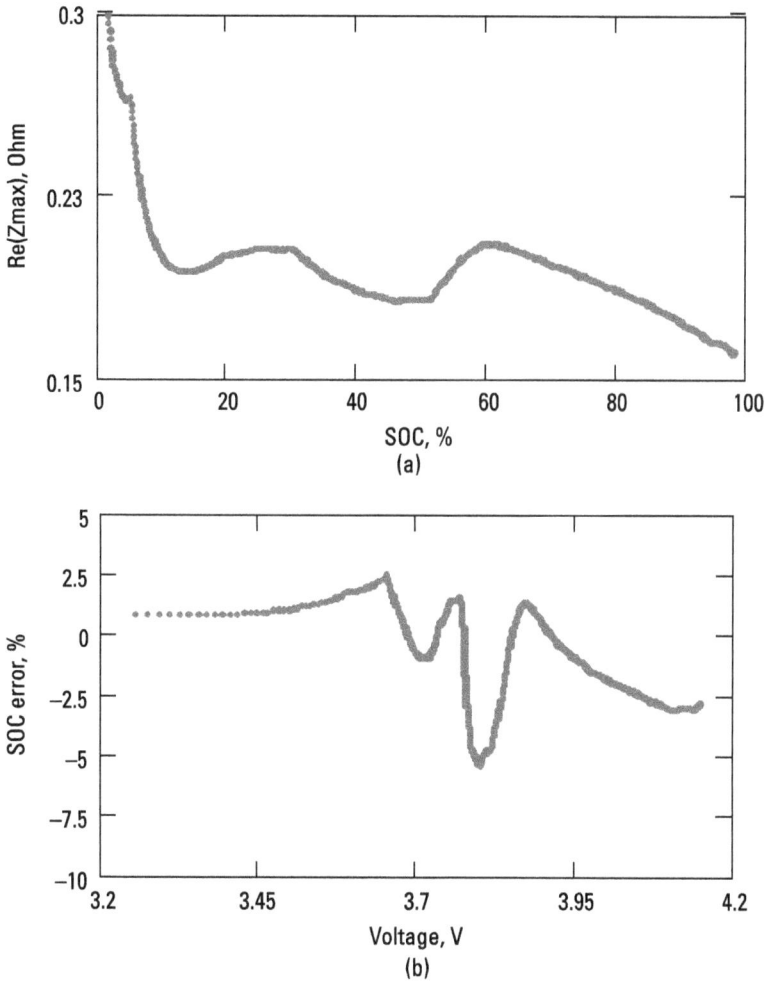

Figure 5.6 (a) DC internal resistance versus SOC. (b) Error in SOC estimation performed for 1/2C rate discharge when IR drop was corrected under assumption of a constant internal resistance (using average value).

high loads. For example, a common 1/2C discharge rate and a typical DC impedance for 2-Ah cells of about 0.15 Ω results in worst case 45-mV differences among cells. This may result in a SOC estimation error of 20%. Figure 5.10 shows impedance spectra for 50 new cells from two different manufacturers, which indicates the usual impedance variations among cells. Note that the high-frequency impedance (left side of Nyquist plot) is very similar for all cells, but the low-frequency impedance (right side of the plot), which actually defines DC performance, shows substantial variation.

Finally, the single biggest impedance-related problem comes when a cell ages. We know that an increase in impedance is much more significant than a

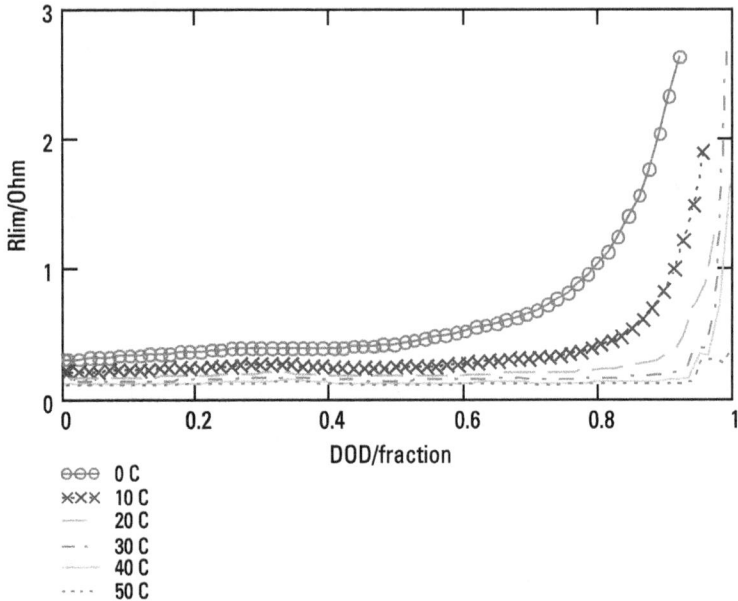

Figure 5.7 Temperature and depth of discharge (DOD) dependency of limiting low-frequency impedance of an 18650 LiCoO$_2$-based lithium-ion battery.

decrease in the full charge cell capacity. A typical Li-ion battery increases its DC resistance (e.g., real part of low-frequency impedance) by about 60% in 100 cycles as can be seen in Figure 5.11. Interestingly, the high-frequency impedance (right side of the Bode plot) barely changes during the same period. Since cell manufacturers usually report only high-frequency impedance, it could give a misleading idea that battery impedance does not change with aging. However, high-frequency impedance is only relevant for short pulses in the millisecond range, but IR drop during persistent discharge (which is relevant for battery capacity gauging) is fully determined by low-frequency impedance. If the effect of the impedance increase is not considered, a voltage-based algorithm that seems to work for new battery packs will fail miserably (50% error) when a pack reaches only 15% of its life estimation of 500 cycles.

The effect of all error contributions is summarized in Figure 5.12. Note that at different SOCs, the error is not going to be the same for a fixed voltage error. That is because the flat portions of the V(SOC) curve have a higher SOC error even if the voltage error is the same.

5.3.3   Hardware Implementation of Voltage Correlation

Simple voltage correlation methods are often implemented on the host microcontroller of a device that uses an embedded battery. Such implementations

(a)

(b)

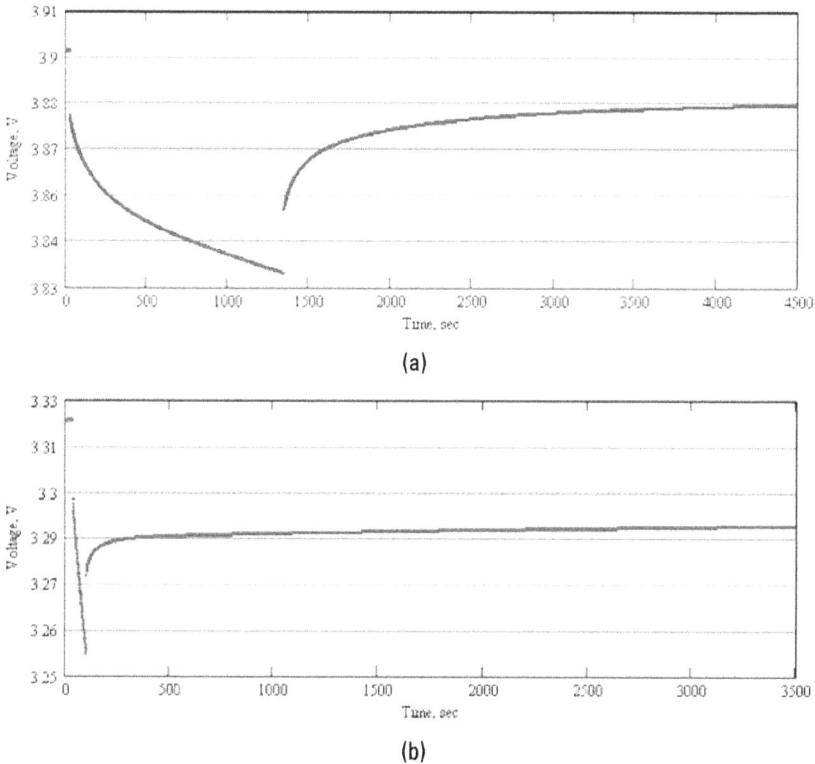

Figure 5.8 Voltage response of a Li-ion battery to an onset of C/2 rate discharge load in the (a) fully charged state and (b) fully discharged state.

use a simple ADC to measure voltage with a 20- to 40-mV error. For the typical V(SOC) curve of the most common LiCoO$_2$-based Li-ion batteries, this transforms into 20% to 40% SOC error in the flat region of the curve. To increase the accuracy of a voltage-based implementation, external voltage monitors can be used, which report accurate voltage measurement results to the host via single-wire communication. One example of such an implementation is the bq2023 voltage monitor. In addition to voltage information, it also provides current and temperature information that can be used to implement IR compensation to voltage for improving accuracy. Its current integration capability also allows us to combine voltage correlations with current integration, as described in the following chapter. Figure 5.13 shows an example application diagram that includes the IC monitor.

5.3.4 Coulomb Counting: Current Integration Method

The current integration method relies on the robust idea that if we integrate all battery charge and discharge currents, we will always know how much cou-

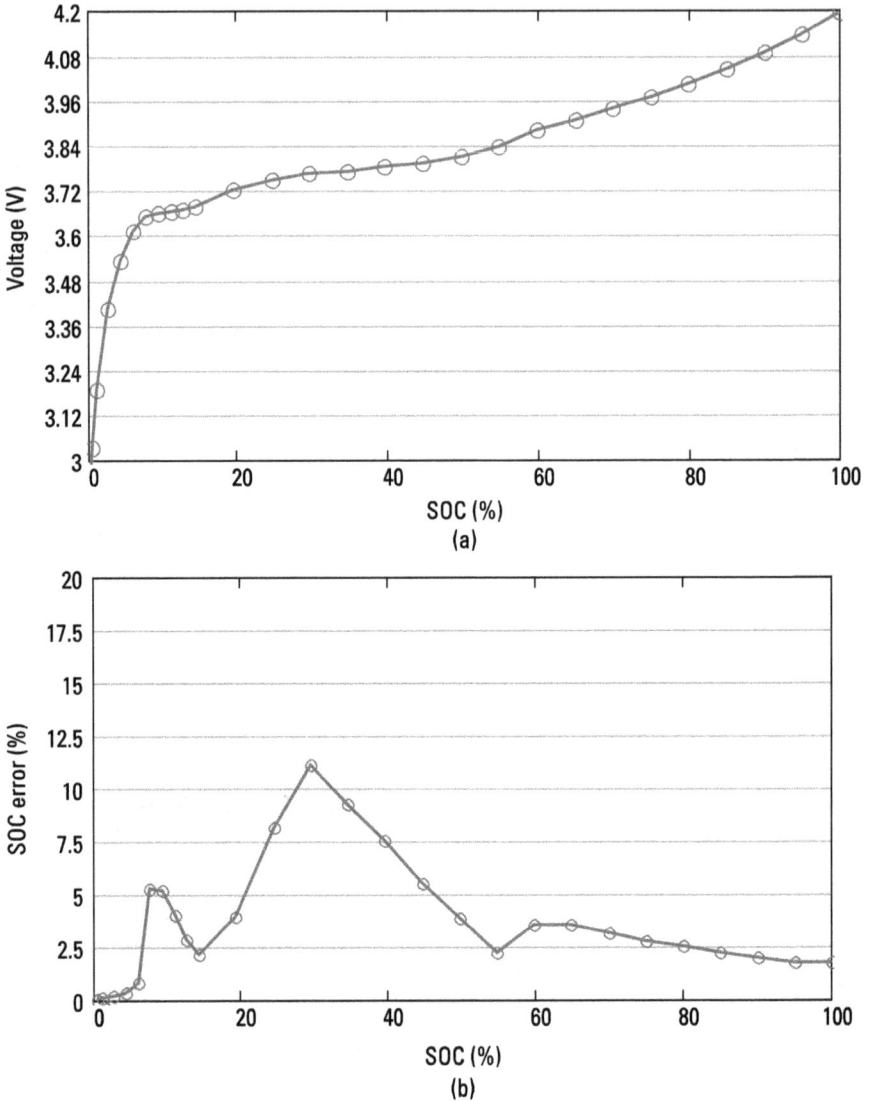

Figure 5.9 (a) Voltage versus SOC profile of a typical LiCoO$_2$-based lithium ion battery. (b) Error in SOC estimation depending on SOC where estimation was performed given a 20-mV error from the transient effect.

lometric capacity is remaining. Integrating the current works particularly well when the initial battery capacity is known and the coulometric efficiency is 100%. In other words, when charging, all of the coulombs that go into the battery stay in the battery and all decreases in battery charge state are due to an external discharge current.

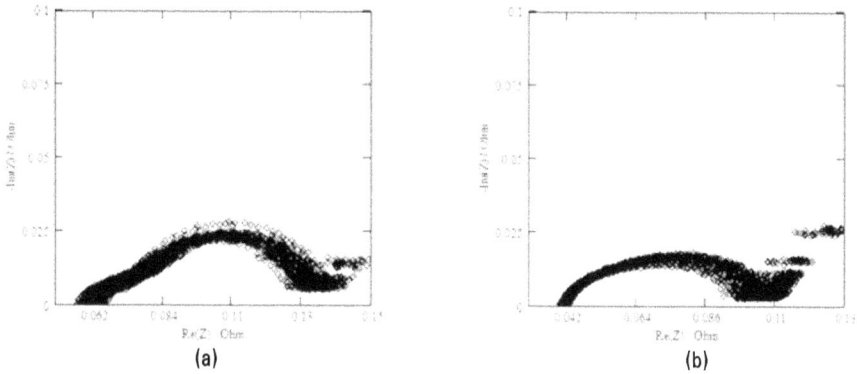

Figure 5.10   Impedance spectra measured at 3.750V from 1 kHz to 1 mHz for 50 new cells from one batch for manufacturers (a) and (b).

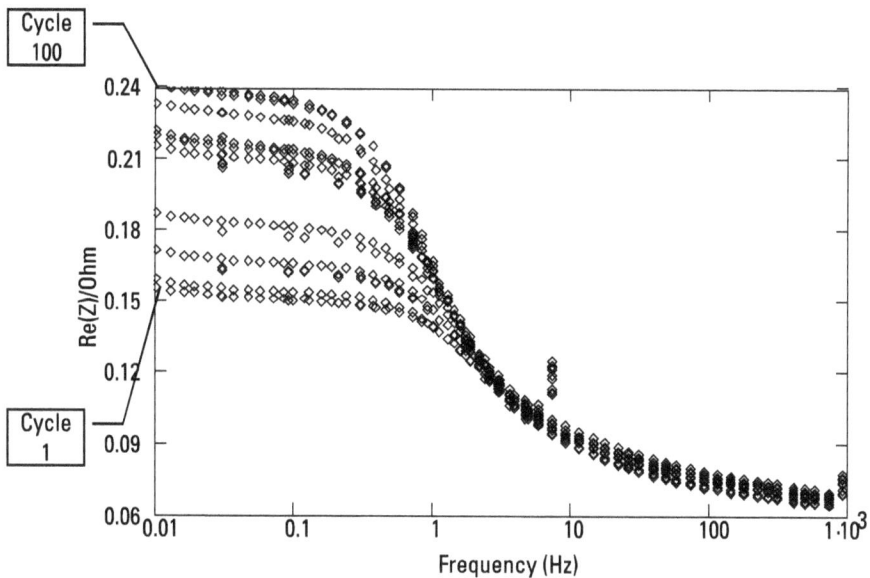

Figure 5.11   Aging effects on a Li-ion battery during 100 cycles. Impedance spectra are in the range from 1 kHz to 1 mHz for cycles 1,10, 20, ..., 100 as shown by a Bode plot [Re(Z) versus logarithm of the frequency].

This seemingly bulletproof approach is modified by predicting self-discharge and battery charging efficiencies. After these modifications the current integration method is successfully used in most currently employed battery gas gauges. However, these modifications to the charge integration methods are principally estimates that may produce errors in particular usage patterns such as those with long periods of inactivity or highly variable discharge current. If the battery is charged and left unused for several weeks or just never fully

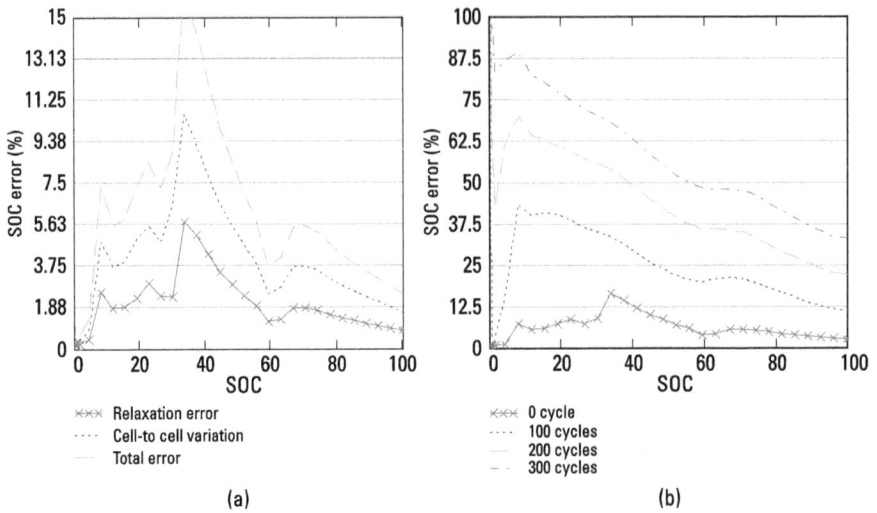

Figure 5.12 (a) Contributions of the relaxation effect and cell-to-cell variations to SOC error. (b) Overall SOC error change with aging.

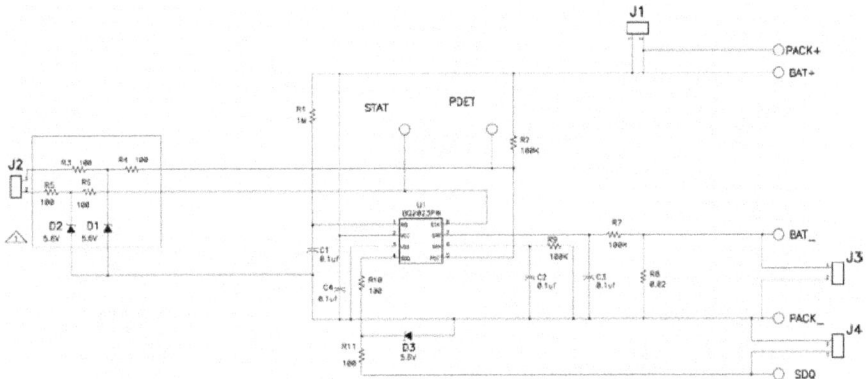

Figure 5.13 Battery pack schematic with bq2023 battery monitor.

charged for many charge and discharge cycles, the self-discharge due to internal chemical reactions eventually becomes noticeable. There is no way to measure the self-discharge current, so it has to be corrected by using a predefined equation. Because different battery models have different self-discharge rates, which also depend on the SOC, temperature, and cycling history of the batteries, exact modeling of self-discharge requires a time-consuming effort in data collection and still remains quite imprecise. Low current drain by electronics and even the battery management system itself is also usually below the detection level of the coulomb counter and will have a significant effect as a "lost" coulomb count

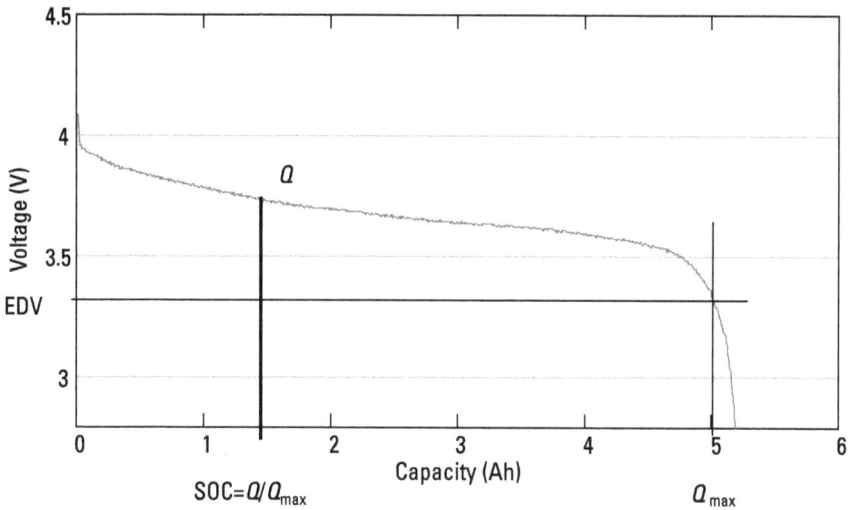

Figure 5.14   State of charge indication using charge integration.

over a long period of time, contributing to error accumulation that needs to be "reset" by reaching a full charge reference point.

Although not restricted to coulomb counting, another problem is that the value of total capacity is updated only if a nearly full discharge occurs very soon after a full charge. If full discharge events are rare or comparatively rare, a considerable decrease in the actual available capacity can commence before its value can be updated by the gas gauge. This will result in overestimation of available capacity during these periods. In some applications (such as laptops) the operating system will terminate the discharge at about 3% remaining capacity to prevent a system crash and to save the data, so a learning opportunity for updating the total capacity will never occur if a full discharge is required.

5.3.5   Coulomb Counting with Voltage-Based Early Learning

To solve the problem of rare or never occurring full capacity learning, voltage modeling can be used. Voltage modeling allows the system to learn the total capacity at voltage thresholds that are higher than the minimal voltage. For example, in the bq2060 gas gauge, fixed voltage thresholds that correspond to 7% and 3% remaining capacity were used, as shown in Figure 5.15(a). This method works well if discharge is always taking place at the same rate of discharge and temperature. The problem with all of these changing conditions can be seen in Figure 5.15(b).

When voltage thresholds are calculated for current I_1, learning correctly occurs at 7% SOC. However, when the current changes to I_2, the same voltage corresponds to about 30% SOC. But the gas gauge is going to assume that this

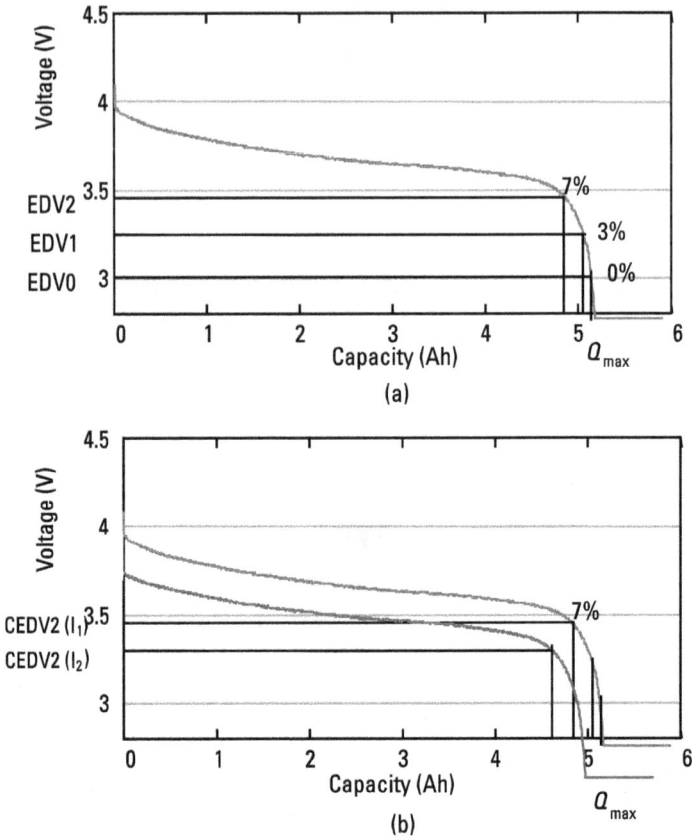

Figure 5.15 Capacity learning before reaching the termination voltage with (a) fixed EDV thresholds and (b) rate-adjustable thresholds.

is 7% SOC when voltage is reached! So it will underestimate the total capacity by 23%. To solve this problem, newer gas gauges starting with the bq2060 are using *compensated EDV* thresholds, or CEDVs. Voltage threshold for 7% is expressed by the internal function V(7%, T, I) = OCV(7%, T) + $I * R$(7%, T). Parameters of this function are battery specific and have to be generated by testing each battery at different rates and temperatures. This improves the accuracy under variable discharge rate conditions for new batteries. However, the resistance function R(7%, T) remains static, but the actual cell resistance changes with aging. So the CEDV function will eventually be out of date and the error will increase. Newer devices such as the bq2084 have a correction factor (called the A0-factor) that can increase resistance R with increasing cycle number. This constitutes some improvement but because battery aging is unpredictable (there is cell-to-cell variation even under the same conditions) and depends on other factors such as temperature, time, and charging conditions, the cycle number is not a perfectly accurate way to adjust resistance.

In addition to inherent inaccuracy, creating correction parameters requires data collection for the voltage profiles at several rates and temperatures for large cycle numbers (up to 500 cycles).

5.3.6 Hardware Implementation of Coulomb Counting Gauging

Current integration is typically done by means of measuring the voltage drop across a sense resistor (R1), as shown in the sample implementation schematic of Figure 5.16.

In implementation of a current integration fuel gauge, the accuracy of the current measurement and integration plays a critical role. Sources of integration error are as follows: gain error, ADC resolution error, offset error, clock error, and sampling error. The gain of the integrating ADC is typically calibrated at the board level to account for sense resistor tolerance. To reduce power losses on sense resistors, small values of 5 to 20 mΩ are typically used. With such a small value, the resistance of board traces plays a significant role, so it is important to provide a kelvin connection (two reference wires) from both sides of the sense resistor directly to the ADC input in addition to the current-carrying traces leading from the cell to the sense resistor and from the sense resistor to the pack

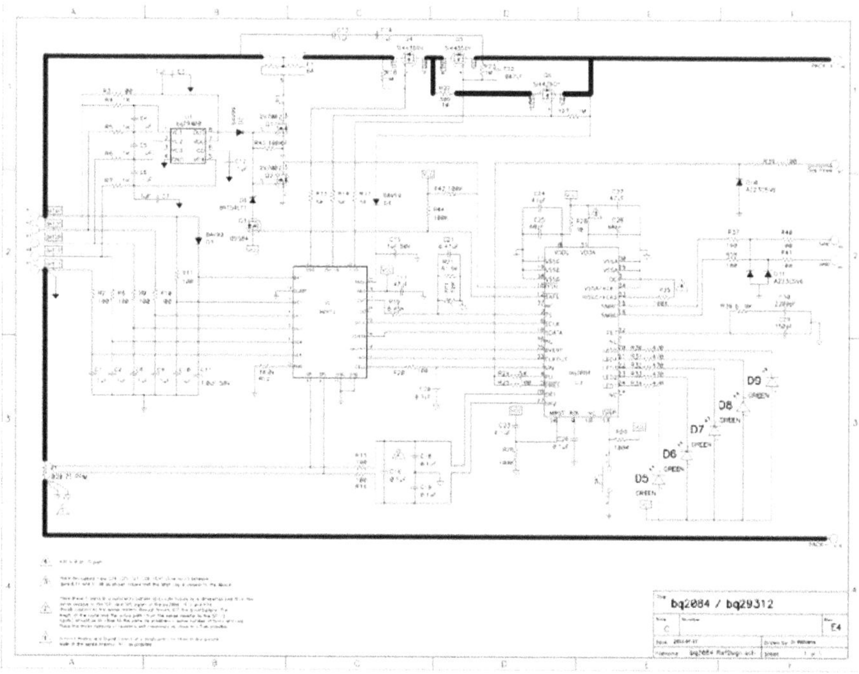

Figure 5.16 Example of current integration with voltage correction (CEDV) in a gas-gauge implementation (bq2084) including an analog front-end (bq29312A) and protection FETs.

terminals. Sense resistors can become hot during active device operation, so we recommend the use of only those resistors with a low thermal coefficient. Due to a very small voltage drop across the sense resistor, noise from other components on the board can contribute a significant distortion. Because of that, use of a filter (C16, C18, C19, R15, and R16) is recommended.

The ADC used for current integration should have sufficient resolution to reduce granularity error (the device discussed above uses 16-bit ADC). Sampling errors can degrade accuracy if the sampling rate of the ADC is not sufficient to capture all load pulses. Oversampling integrating sigma-delta converters or voltage-to-frequency converters (VFCs) can provide adequate integration for most applications.

While simple current integration requires only a crude voltage measurement to detect end of discharge, the schemes that allow learning without reaching the end of discharge (EDV or CEDV) require a 20- to 40-mV maximal voltage error limit for accurately detecting the learning threshold.

Using adjustable thresholds and other sophisticated logic requires the use of a microcontroller as part of the capacity indicator. It is also preferable to have programmable memory on board so that thresholds calculated specifically for a particular battery model can be programmed during production. Programmable memory also allows for increased flexibility for supporting various safety settings, cell balancing settings, and so forth.

5.4   Advanced Gauging Methods: Impedance Track™

5.4.1   Basic Concept

To improve the prediction accuracy of remaining capacity, a combination of voltage and current integration can be used. Although EDV and CEDV methods use this combination near the end of discharge, the application of these two methods at all states of charge provides important benefits, especially for those batteries that have capacities that vary greatly with discharge rate and temperature.

In this book we use the TI Impedance Track™ gauging method as an example of advanced voltage and current-based battery gauging partly because it is indeed the most commonly used advanced gauging method in the world today and because it is one of the few that is fully disclosed in patent literature so no black box walls obscure its operation.

The basic principle of Impedance Track (IT) is simple: the operation of a device is monitored, and depending on the present mode of operation, the gauging mode that gives the highest accuracy is used. Indeed, most devices typically have a mode of active use, where it is actually being utilized for its purpose (e.g., a notebook being turned on and perused), and a mode of passivity (e.g., when a laptop is shut down). In IT, voltage correlation is used when it works best, for example, during a period of inactivity in which current is very small and so its contribution can be corrected with low error. In contrast, the coulomb counting method is used during periods of active operation, where current integration has the highest accuracy (the higher the current, the less effect the offset error of the coulomb counter has). The complexity remains in how to reconcile these two measurement methods, as well as how to report remaining capacity that is adequate for the present conditions (rate of discharge and temperature). Let's look at different parts of the algorithm a little closer to see how it all fits together.

5.4.2 Voltage Correlation in IT

First let's look at the voltage correlation aspect of the algorithm. Voltage correlation is performed only if the "inactive mode" of device operation is detected, and sufficient time has passed so that all transient effects of the previous load have declined. The gauge first detects that the current dropped below a predefined threshold so that inactive mode entry is confirmed; then it continuously monitors battery voltage until full relaxation is detected as the voltage change becomes slower than 1 μV/s (basically no noticeable change above the noise level). Note that the current does not need to be zero, but just small enough (below C/20 rate) so that the IR correction can be applied to voltage with little error.

Once relaxation is detected, due to very precise correlation between open circuit voltage and SOC, the voltage method allows for a precise SOC estimation. In this way, the periods of inactivity allow us to obtain a very accurate "reference point" not affected by all of the various error contributions common to voltage correlation under load that we discussed earlier, and to get an exact "starting position" for SOC, as indicated in Figure 5.17.

In this manner, the need for self-discharge estimation for the periods of inactivity is eliminated, since the self-discharge contribution is reflected in the OCV. In addition, the inaccuracies due to measuring very light loads similar to those presented by the battery pack electronics and system leakage paths are eliminated.

Because fewer chemistries exist compared to the number of battery models, all characterization to obtain OCV tables can be done by IC manufacturers and so no testing and data collection need to be performed by battery pack designers. If the chemistry to be used is unknown, a selection table or a simple selection test can be used to find the correct chemical ID.

Prior to the device being switched on again, the last SOC reading needs to be retained. Since it is not known when a device will actually turn on, periodic measurements of voltage are made and the last SOC is kept ready for the time

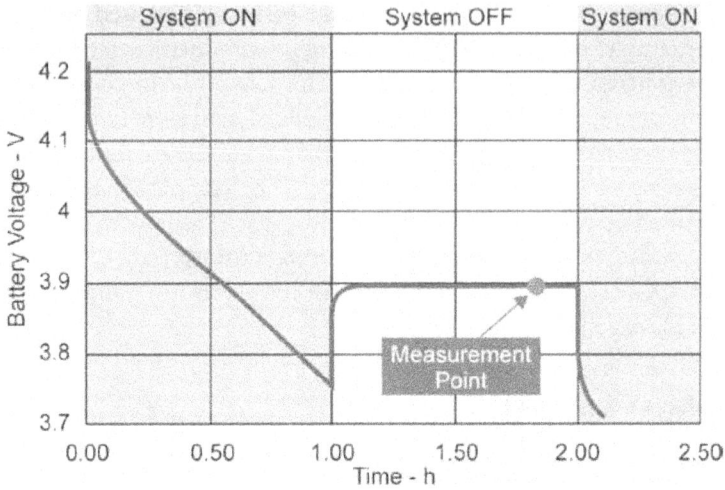

Figure 5.17 Determining the starting SOC for the next active period by OCV measurement after relaxation from the previous active period.

when active operation will start. When the device is switched to the active state and a load is applied to the battery, current integration takes over and starts to correct the last SOC obtained from the voltage correlation. However, we need to convert the coulomb count obtained from current integration (whose units are mAh) into SOC (units %). To get to SOC increment, we need to divide the coulomb count by total capacity: $SOC = SOC_0 - Q_{pass}/Q_{max}$. But where can we get Q_{max}? Do we again need to perform a full discharge to obtain it?

5.4.3   Full Chemical Capacity (Q_{max}) Update in IT

Interestingly, the ability to update the SOC during periods of inactivity in combination with coulomb counting allows for an elegant way of updating the total capacity without full discharge. This is one of the main patented insights behind IT. Indeed, when we have information about the SOC before applying a load, precisely measurement of the passed charge and the SOC after the load is removed as shown in Figure 5.18, we can easily determine total capacity.

For the case of charging [Figure 5.18(a)], the OCV reading is taken at point P1 before charging begins. The voltage correlation allows us to obtain $SOC_1 = f(OCV_1, T)$. Coulomb count Q_{pass} is accumulated during charging. Then after charging is terminated, when a period of inactivity is detected and sufficient relaxation of the cell is reached, a second OCV reading, OCV_2, is taken, and a second SOC value, SOC_2, is obtained (note that for the purpose of these discussions SOC values are expressed as a fraction of 1). Given these values, total capacity can be obtained from its definition as $Q_{max} = Q_{pass}/$

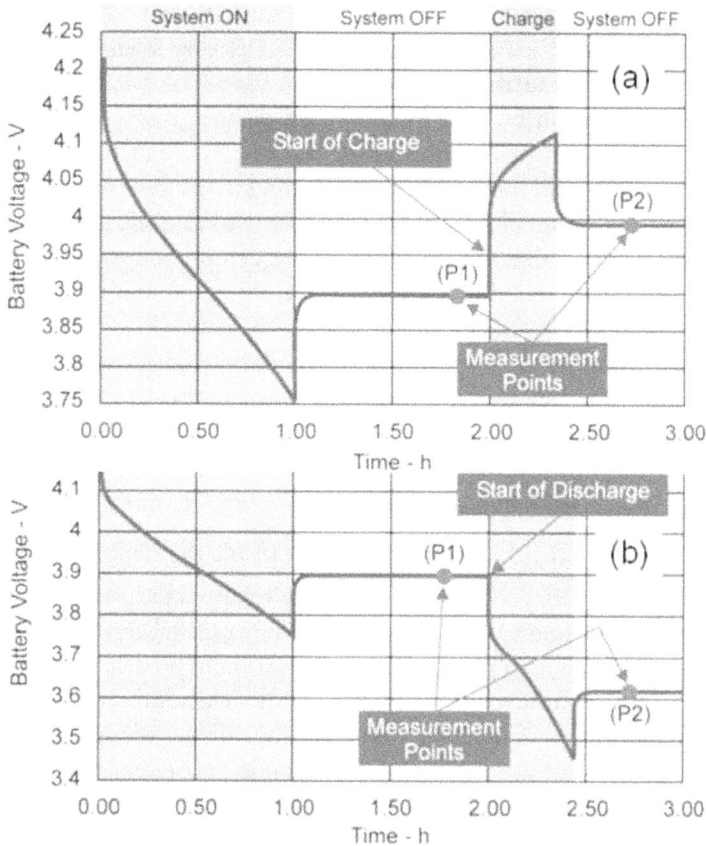

Figure 5.18 Determining total capacity by using SOC points before and after the (a) charge activity and (b) discharge activity.

$(SOC_2 - SOC_1)$. An analogous method works for the discharge case, as shown in Figure 5.18(b).

One amazing result of this simple relationship is that learning the total capacity of a battery no longer requires a full charge or a full discharge. It can be learned with any starting SOC point and any ending SOC point! This is a huge advantage to systems that are very rarely fully discharged, which happens to be most consumer portable devices, such as laptops, cell phones, and tablets, as well as a range of other battery-operated systems such as the recently popular hybrid electric vehicles. One especially curious case when this operation is absolutely necessary is found in systems where a full discharge is not only unlikely, but impossible due to system needs. An example of such systems are server and other essential infrastructure backup batteries that can never be discharged because this would remove their protective function. However, they can be partially discharged by a small amount that still ensures required backup time, yet allows them to update their usable capacity and to verify that their "state of health" (see discussion below) is still satisfactory to ensure the required backup time will be available when needed.

This method can be used whenever the determination of the SOC and measured capacity removed is known to be accurate. Cases such as small changes in SOC or capacity will lead to an elevated error, but limits of minimal required SOC change and minimal passed charge can be used to disqualify an update in such situations. Note that the total capacity we determine by this method corresponds to "no load" conditions, for example, the maximum possible capacity that can be extracted. Under a nonzero load, capacity will be reduced due to IR drop, which causes the termination voltage to be reached earlier under load. Determining the usable capacity and what is needed to achieve it is discussed in subsequent sections.

5.4.4   Battery Impedance Update in IT

While the question of chemical SOC is solved by the preceding approach without considering the effect of cell impedance, cell impedance is still an important parameter for adjusting total capacity for the rate of discharge. Indeed the device user and a system that uses gauging information to predict when it needs to go to safe shutdown usually do not care about the chemical state of the battery as such. What they care about is when exactly will the voltage go below the termination voltage of this particular system under its present load and present temperature. This moment has nothing to do with chemical SOC = 0, which is usually never reached in real devices because their IR drop causes the cell voltage to reach the termination voltage much earlier and some chemical capacity is still left at that moment. This means that the value for the SOC that should be reported to an end user that would make sense would show zero not when the chemical state is zero, but when the termination voltage is reached. To differentiate from the chemical SOC, this value is called the relative SOC (RSOC) in smart battery bus (SMBus) specification adopted by most notebook systems. This is also the value reported by Impedance Track gauges, while the chemical SOC information is kept internally to the algorithm.

If the cell impedance dependencies on chemical SOC and temperature are known, it is possible to employ simple modeling to determine when the termination voltage will be reached at a given load and temperature. However, as mentioned before during the discussion of different contributions to voltage correlation errors, the impedance value has a large cell-to-cell variation even in the same batch and increases rapidly with cell aging and cycling, so it would not be sufficient to just store it in a database.

To solve this problem, Impedance Track uses real-time impedance measurements, which can keep an impedance database continuously updated.

Impedance measurements are made by comparing the open circuit voltage for a given chemical SOC from the OCV table with a presently measured voltage as shown in Figure 5.19(a). The resulting impedance dependence on SOC is shown in Figure 5.19(b).

Such real-time impedance updates eliminate the problem with impedance variations between cells and cell aging. Of course, in a real system the discharge rate is rarely constant. See, for example, voltage and current variations recorded in a notebook battery during typical usage in Figure 5.20.

Each load variation creates a temporary transient effect, which would cause an error in the low-frequency impedance measurement if we ensure that the voltage difference between the OCV and the present voltage does not have a transient effect and is fully stabilized. These variations of real-life loads contribute significant complexity to the measurement. The IT algorithm continuously

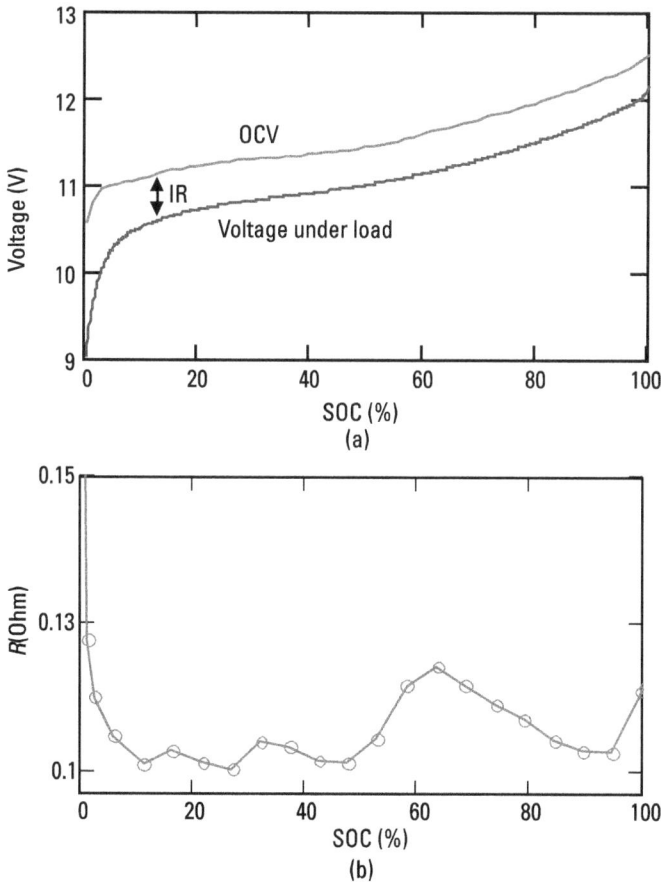

Figure 5.19 (a) Open circuit voltage and voltage under load. (b) Impedance curve obtained from comparing two voltage curves.

(a)

(b)

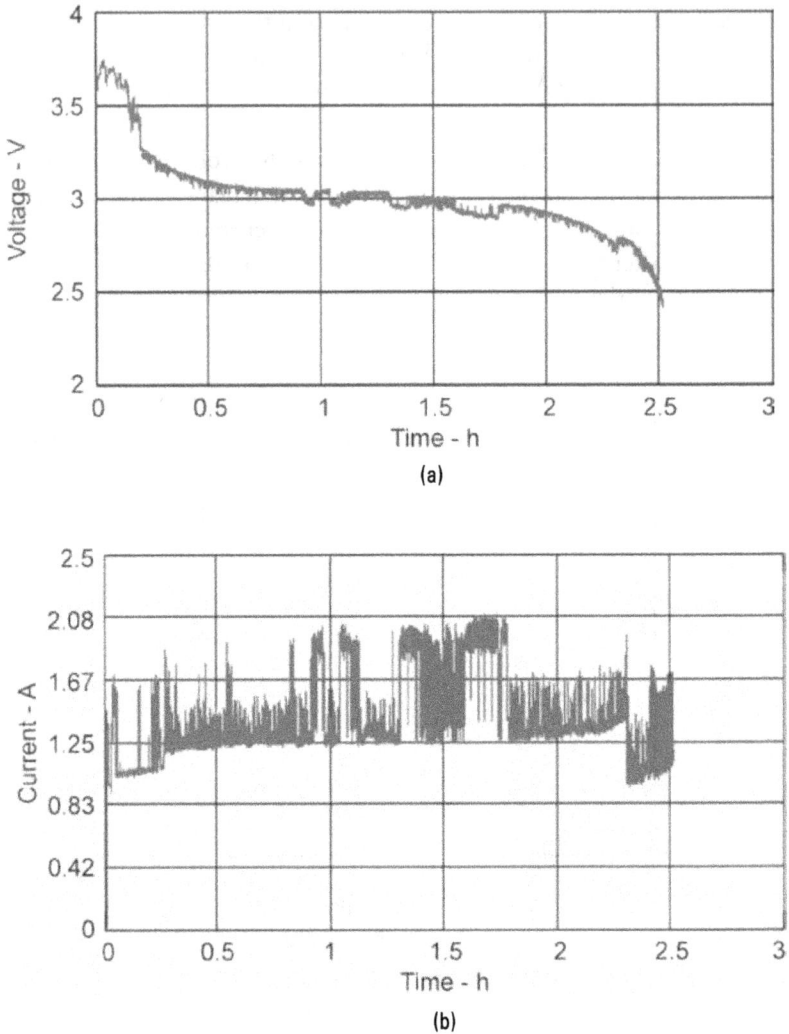

Figure 5.20   (a) Voltage and (b) current variations on a notebook battery during typical usage.

monitors load variations and detects when a transient effect occurs to filter it, compensate for it, or disable the update altogether depending on particular conditions.

5.4.5   Thermal Modeling to Account for Temperature Effects on Usable Capacity

Now that we have up-to-date impedance values as well as the temperature dependence of impedance (as shown earlier in Figure 5.7), can we now model the voltage profile during discharge? As it turns out, not quite. While we know how to change impedance with temperature, $R(T)$, we do not yet know T itself.

This is because when a device starts operating, the battery temperature is going to increase due to self-heating as can be seen in Figure 5.21.

We need to predict the remaining capacity both at the beginning of device operation (before it started to self-heat) and during operation. Remaining capacity depends on the point at which the voltage under load will reach the termination voltage, which will happen far in the future where temperature is already going to change significantly. This means that we need to substitute into $R(T)$ to find this termination point, which is not the present temperature, but the *future temperature*. This temperature modeling need appears theoretical at first, but it has dramatic effects at low temperature where battery impedance is very high. If not for self-heating, most Li-ion cells would not have any capacity at all at low temperatures, which becomes evident if you just substitute $R(T = -20°C)$ and observe the impedance of tens of ohms, which would cause voltage to instantaneously drop to below the termination limit at any reasonable load. However, due to self-heating, the starting temperature is likely to increase by 15° to 20°C and so by the time termination is reached the battery is already at a balmy 5 to 10°C, its impedance is low, and its voltage stays above the termination point. So self-heating makes a difference between having almost full capacity or not having any capacity at all.

To predict the future temperature, Impedance Track implements thermal modeling that considers the thermal capacity of the cell, the temperature of outside air, the heat exchange coefficient between the battery and the outside system, and the heating rate. All parameters of the model are obtained by observing the actual temperature change of the system automatically, so the

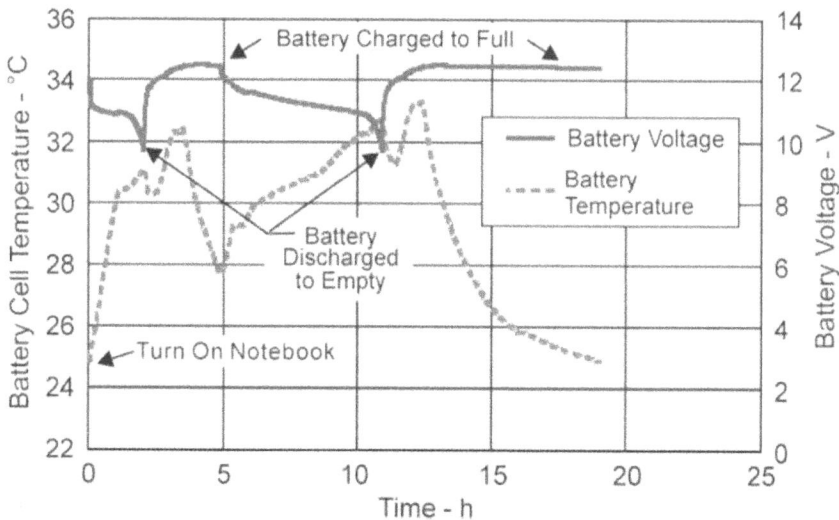

Figure 5.21   Battery temperature and voltage profiles during notebook PC operation.

model remains accurate even if heat exchange conditions change; for example, a notebook is moved from a table-top (high exchange rate) to a pillow (very heat insulating). The largest part of the heat generation in a cell is due to its internal resistance, $I^2R(T)$, but resistance itself depends on temperature. That makes the model strongly nonlinear and only iterative calculations of temperature are possible. But as the end result, such a self-updating thermal model allows us to determine not just the present temperature, but the temperature at a time when termination will occur, ensuring the correct estimation of remaining capacity.

5.4.6 Load Modeling

Given all of the necessary machinery for modeling, we will also have to decide how optimistic or pessimistic we want to be about assuming what load the system will experience close to the end of discharge. Because the load can change (the load does determine how high or low the IR drop close to termination is going to be), there is some inherent uncertainty in predicting remaining capacity especially if it has to be done a long time beforehand. However, this uncertainty usually decreases when we have some time to observe the system. For example, when a user is just turning on a notebook PC, it is almost impossible to predict what the load is going to look like at the end of discharge. But once we observe what the user is actually doing, we would be able to differentiate, for example, between typing in a Word document or playing a video. This is done by observing the average system current (or power, which is more accurate). The total average current from beginning of system usage is very likely to be representative of what will happen at the end of discharge. But if the load is going to change suddenly (for instance, if the user switches from Word to playing a game), it is desirable to detect this load change and to start using a new load model. Load modeling is in a way a game of detecting human behaviors and because of that it will never be perfect, although it can gradually improve. It can usually be more accurate if more is known about the nature of the system. Because each system designer knows a lot about his or her system, it makes sense to allow the designer to influence how the load will be modeled. For this reason Impedance Track includes many configuration options defining different ways of load measurement and averaging. Choices include using an instantaneous load, the total average of the present discharge, or a 12-second time constant running average [known as the smart battery specifications (SBS) average], as well as user-selected load and more sophisticated choices that give a conservative estimate such as the largest 12-second average ever observed during discharge.

Finally, it makes sense to say a few words about choosing to model constant current or constant power. It makes a big difference, because in the constant current case, the present current will be measured and extrapolated to be the same until the end of discharge. However, most electronic systems have a DC-DC converter that converts variable battery voltage (which decreases with battery discharge level) into a constant voltage that powers the system. The result of such conversion is that while the system is drawing constant current at the constant voltage of the converter's output, the battery is experiencing a current increase when its voltage is lower, so it is operating under constant power conditions. Given a compete voltage and impedance model of a battery as described above, it is easy to model voltage response for either constant current or constant power conditions. Impedance Track offers both options. For portable electronic systems constant power is the best option, whereas for some systems without a DC-DC convertor such as a portable drill constant current could be adequate. Note that time-to-empty is also much better calculated by dividing the remaining energy (Wh) by the average power consumed (W), rather than using remaining capacity in Ah and current consumed—because the current will actually increase, making the calculation incorrect. Impedance Track will calculate time-to-empty according to the energy or capacity reporting choice selected by a configuration bit defined in the SBS, so it is an advantage to select energy reporting.

5.4.7   Bringing It All Together: Predicting Usable Capacity and Energy for Present Conditions

Bringing together all of the ingredients we collected in the previous sections, we now have a recipe for correctly predicting remaining capacity and energy under any temperature and load conditions. Let's recap all of the ingredients:

Battery Information
- Open circuit voltage dependency on chemical SOC.
- Open circuit voltage dependency on temperature.
- Resistance dependency on chemical SOC.
- Resistance dependency on temperature.
- Temperature change modeling parameters.

Measurements on the System
- Voltage.
- Current.
- Present temperature.
- Integrated charge, Q_{pass}.

Derived Values

- Chemical SOC.
- Updated total chemical capacity, Q_{max}.
- Updated resistance dependency on chemical SOC.
- Updated thermal modeling parameters.
- System load chosen by load model.

Once all choices have been made, it is straightforward to model battery voltage as

$V(SOC, T) = OCV(SOC, T) + I * R(SOC, T)$. Iterative computation of this equation, which stepwise decreases the SOC by a small increment from a preset state until $V(SOC, T)$ < termination voltage is reached, allows us to find the chemical SOC that corresponds to the termination point, SOC_{final}. Given the chemical SOC at termination, and knowing the total capacity Q_{max}, we can then easily calculate remaining capacity as $Q_{rem} = Q_{max} * (SOC_{present} - SOC_{final})$, where SOC is represented by a unitless fraction of 1. Similarly, remaining energy is found by integrating all voltages and current during each iteration step. Note that this energy computation is much more accurate than just taking mAh capacity and multiplying it by some kind of average voltage, because the effect of the IR drop on energy is correctly accounted for.

A constant power computation can be performed in a similar manner except that current I is also becoming a function of SOC since the requirement power = $I * V$ has to be maintained, which makes computation more recursive.

As can be seen in Figure 5.22, the preceding computation, which uses data from a self-updating database, enables remarkable precision for voltage profile predictions at any given load.

In most cases an error in usable capacity estimation below 1% can be achieved and, most importantly, high accuracy is sustained throughout the entire life of the battery pack. Here is the result of the recipe execution:

- Remaining charge at present conditions, mAh.
- Remaining energy at present conditions, mWh.
- Total capacity at present conditions, mAh.
- Total energy at present conditions, mWh.
- Relative state of charge, RSOC (reported either as remaining charge * 100/total capacity or as remaining energy * 100/total energy, depending on SBS capacity mode bit).
- Remaining run time.

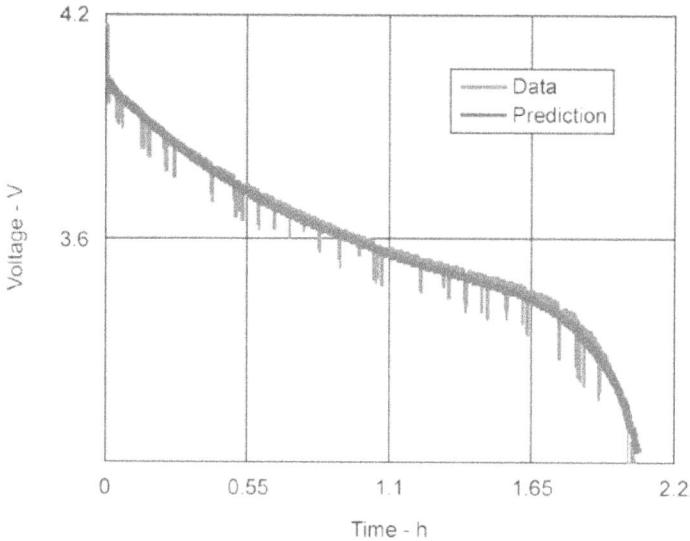

Figure 5.22 Voltage profile predicted by fuel-gauging algorithm on the basis of real-time updates of cell impedance versus a voltage profile subsequently measured using typical notebook load experimental data.

Using these values to control the system's soft shutdown at the point at which only energy required for the shut down is left will allow us to achieve the optimal result of providing the end user with the maximal run time, while ensuring that data are preserved and no unexpected shutdowns occur.

5.4.8 State of Health

While reporting the amount of remaining energy under present conditions is important with regard to actual system control, it has one disadvantage: It will vary with load and temperature, so it cannot be used to judge whether your battery itself is in good shape. For example, your new cell was delivering 1,000 mAh, but if tested under 0°C, the same cell will deliver only 600 mAh. That is what Impedance Track gauge will also predict (correctly). However, if you use this value of capacity to evaluate the health of your battery, you might wrongly conclude that your battery is degraded by 40%! The same issue will happen if you evaluate your battery at a low rate of discharge (say, C/10) and then test the capacity at a high rate (say, 1C rate). Again it will look like the battery has degraded, even though it would again deliver the same capacity as before if again discharged at the C/10 rate.

To avoid such false estimates, the right way to evaluate battery degradation would be to compare its capacity under the same conditions. For example, you could compare the capacity you obtained for a new battery at standard conditions (25°C temperature and C/3 rate), that is, the average rate for your system, against the capacity you obtained for an aged battery under the same conditions. In this case, the ratio between aged capacity and new capacity would give you a meaningful indication of state of health. The only problem with this approach is that you cannot ask a user of, say, a laptop to run a discharge test at precise, fixed temperature and discharge conditions. The closest users can come to a "constant temperature" in their home environment would be by placing a laptop into a refrigerator or into a baking oven, which both appear to be a very bad idea for different reasons.

Fortunately, the Impedance Track gauge can *model* the discharge of a battery under any conditions. So instead of requiring all notebook or cell phone users to purchase a thermal chamber, we can just algorithmically set a value of temperature = 25°C and load = C/3 load and run a simulation that will produce total capacity at the reference conditions every time when either capacity or impedance has changed. Using this *reference total capacity*, we can report the state of health (SOH) of a battery as a ratio of a present reference total capacity to the new battery reference capacity (usually stored as design capacity) at any time without requiring any actions on the user side.

State of health information is useful to device end-users to determine when they have to replace the battery. Traditionally in the battery industry a battery is considered dead at SOH = 70% (usually reached after about 500 cycles). The reason for this is that after 70%, a battery's usable capacity tends to start decreasing very rapidly and quickly goes to zero. It happens because at 70% usable capacity the IR drop is so high that the voltage curve under load is starting to "touch" the termination voltage of the device with its flat portion rather than terminating in a steep portion of the curve as a new battery does. In the flat portion, even a slight further increase of the IR drop causes very large differences is usable capacity. Basically when a battery is at this point, run time becomes very unpredictable with load variations and temperature such that the device can no longer be considered reliable. Also soon after that, the capacity will very quickly move toward zero so it is a good indication that a new battery is needed.

In the case of backup systems, state of health is even more critical because certain backup time needs to be guaranteed so that data loss is prevented. Because backup batteries are always in a fully charged state, SOC as such is not of any interest, but SOH is of great value because the required run time can be expressed in terms of minimal SOH. (Of course, some margin would be added to it to account for the time it will take to replace the battery after the subcritical SOH is reported.)

Note that due to the deep discharge requirement of coulomb counting gauging methods, Impedance Track is the only method that allows SOH to be updated to reflect the currently available usable energy and capacity without taking the backup battery out of operation. As for voltage-based methods, they report only SOC from voltage correlation without any information about actual usable capacity. A voltage-based gauge will report, say, 30% but it not clear of what—is it 30% of new 1,000-mAh battery capacity or of fully degraded 100-mAh capacity? For this reason voltage correlation methods cannot report SOH inherently under any conditions.

5.4.9 Hardware Implementation of IT Algorithm

Hardware implementation of the Impedance Track algorithm is similar to that for coulomb counting gas gauges. Special attention needs to be given to voltage accuracy of the measurement system, because 1 mV in voltage error in the worst case (flat portion of the voltage curve) translates to 1% of SOC error. Calibration routines are included as part of device firmware for all IT devices, which makes calibration in a production line straightforward.

In contrast, the current integration error becomes less important because current measurement errors are compensated for by updating resistance tables and Q_{max}. Although absolute measurement of mAh will be still affected by measurement errors, relative values, such as SOC, and remaining run time will be unaffected. This is especially important for applications that have long periods of inactivity where the offset error becomes significant. The charge integration error accumulated due to ADC offset is canceled through new SOC estimations from the OCV measurement, which allows us to relax certain requirements on the hardware and, therefore, reduce its cost. For example, a smaller sense resistor can be used without changing the ADC. In some applications, current gain calibration can be eliminated completely.

The IT algorithm can run on microcontrollers similar to those used to implement the above-described CEDV gas-gauging algorithms. However, being a newer algorithm it usually runs on ICs that are more modern, for example, ones that combine digital (gauge) and analog/protector (AFE) parts in a single package. This significantly reduces the solution's size.

Flash memory is required to enable support of different battery chemistries, because OCV tables will change if, for example, instead of $LiCoO_2$ a different cathode material is used.

Examples of impedance track gas gauges are the bq20z45 and bq30z55 for notebook implementations, and the bq27541 pack-side gauge for cell phone and tablet computer implementations.

5.5 Host-Side and Pack-Side Gauging

Most of the preceding discussion assumed that a capacity monitor can be continuously connected to the battery. However, many low-cost applications have battery pack electronic components that have been simplified to the minimum needed to provide basic safety such as short-circuit protection. However, on-board the host device, a microcontroller is usually present that could have sufficient processing power to handle capacity monitoring. What issues would arise if we do that?

The first issue is firmware development and handling of capacity monitoring algorithm implementation. The host microcontroller development team typically has no experience with handling capacity monitoring and have no knowledge of battery behavior, the learning curve can often be prohibitive, especially if many different models have to be used and data acquisition is required for each model support. Additional issue arises because any accurate measurements have to be interrupt driven and adding interrupts to the main firmware can often interfere with the main application activities with which the host is involved. This prevents segmentation of firmware development and often creates a firmware development nightmare, which can be easily resolved by using a separate IC to do all of the interrupt-intensive monitoring work.

Measurement accuracy can also be an issue. Although it is possible in principle to use the general-purpose ADCs present on host microcontrollers for conducting current and voltage measurements, the typical accuracy of such devices allows for about a 25% SOC measurement error even without any method-related errors. Added errors from battery aging often makes such measurements too inaccurate for data-processing applications (PDAs, high-end phones) where a soft shutdown is used to preserve data. Use of an external measurement monitor IC that communicates over a single-wire interface with the host and reports accurate measurements (such as bq2023) can provide assured measurement accuracy as well as firmware development support from the IC developer.

Regardless of host micro or external micro implementation, the choice of the method is also critical for this application. Current integration methods have the problem that current integration information will be lost if the battery is disconnected from its monitor and powered down. This makes a direct current integration method only possible if the host itself is able to initialize the capacity to a reasonable value after disconnect.

Overall, voltage correlation methods (while being inherently less accurate due to IR-drop uncertainty) are more suitable for this application because the SOC can be reinitialized after reconnection of the battery. The IT algorithm is especially suitable because it combines the ability of voltage-based algorithms to be initialized to correct the SOC after disconnect with the high accuracy of current integration gauges. A schematic of the typical host-side gauging IC implementation is shown in Figure 5.23.

More complex issues arise if more than one battery has to be used with one device. In this case even if the SOC is correctly initialized, other battery-specific information, such as cell impedance and total capacity, will still be uncertain (assuming only cells with the same chemical ID are allowed). This information

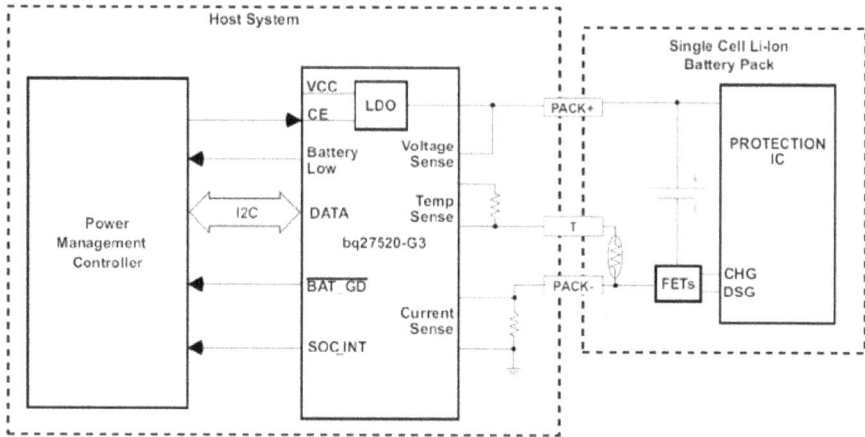

Figure 5.23 Typical application of a host-side gauge IC exemplified by a bq27520 host-side gauge.

can be relearned in the process of further usage, but if packs are interchanged on a regular basis, the overall accuracy will be affected. Placing data-flash on the battery pack side allows for the storage of all cell-specific information or at least for the identification of information sufficient to determine which of the two existing packs is inserted now. Even in the absence of such an ID, a gas gauge can employ some fast measurements that will allow such a determination. However, the number of battery packs routinely used with one device would be limited by the memory dedicated to store all pack information in the host.

Impedance Track gauges especially designed for host-side implementations include the bq27505 and bq27520 gauges.

5.6 Summary

We have reviewed current integration and voltage correlation methods for capacity monitoring. Voltage correlation methods are simpler to implement and are suitable for host-side implementation but have lower accuracy due to various error terms associated with cell impedance compensation. Current correlation methods are accurate and can be successful in applications that often have a full discharge for recalibrating capacity. Care has to be taken in the design of such a monitoring device to provide accurate current integration. For devices where a full discharge is not possible or rare, additional voltage-based prelearning methods have to be used to update to full capacity without full discharge. They do require data collection that is cell specific, as well as more sophisticated hardware including a microcontroller. Use of a combination of voltage-based and current-based methods such as the Impedance Track algorithm allows us to reap the benefits of both methods while eliminating the aging effect on accuracy. Due to its self-updating nature, data collection on the battery pack maker's side is also eliminated, which results in a shorter design time. Accurate host-side or pack-side implementation of battery-gauging ICs is possible.

It looks like the issue of reporting the right number of miles a car will travel regardless of road conditions is finally resolved. What is left is to move the gauging technology from battery-operated devices to cars or, alternatively, to move cars from using gas engines to using batteries.

6

System Considerations

6.1   Introduction

In terms of a battery management system, the system-level electrostatic discharge (ESD), electromagnetic interference (EMI), and other specific design considerations are the main challenges for optimizing the system-level performance. These issues can be extremely challenging and time consuming because they are usually addressed at the end of the design process. Once one of these issues occurs, the circuit board may need to be designed to improve the system-level ESD, EMI, or thermal issues. In this chapter, we briefly analyze the system-level ESD, EMI minimization, and specific design considerations for each piece of end equipment.

6.2   Battery Pack Electronics: General Considerations

Figure 6.1 shows a typical battery pack block diagram for a multicell battery pack, which consists of a fuel gauge IC, AFE circuit, and independent second-level safety protection circuit. The fuel gauge circuit is designed to accurately report the available capacity of Li-ion batteries. Its unique algorithm allows for real-time tracking of battery capacity change, battery impedance, voltage, current, temperature, and other critical battery pack information. The fuel gauge automatically accounts for the charge and discharge rate, self-discharge, and cell aging, resulting in excellent gas-gauging accuracy even when the battery ages. For example, a family of patented Impedance Track™ (IT) gas gauges such as the bq20z70, bq20z80, and bq20z90 can provide up to 1% gauging accuracy over a battery's lifetime. A thermistor is used to monitor the Li-ion cell

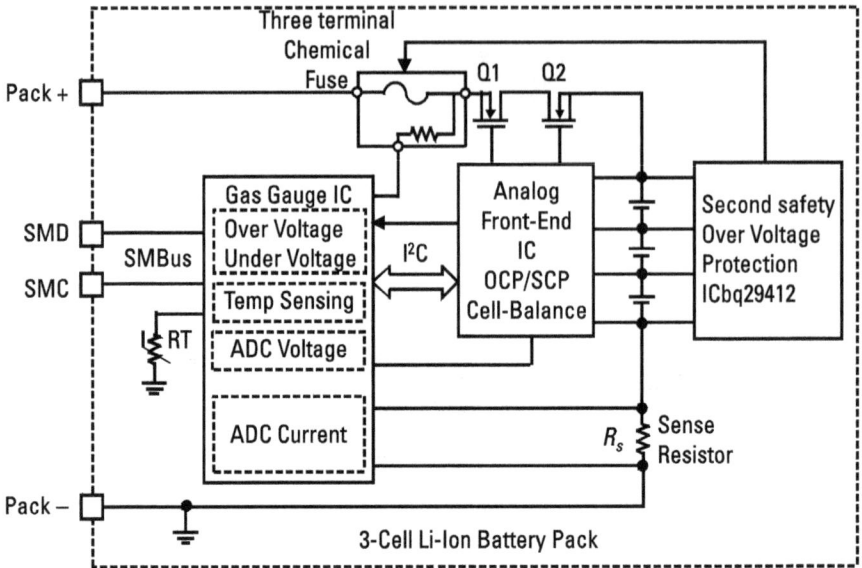

Figure 6.1   Battery management electronics block diagram.

temperature for cell overtemperature protection and for charge and discharge qualification. For example, the battery is usually not allowed to charge when the cell temperature is below 0°C or above 45C, and is not allowed to discharge when the cell temperature is above 65°C. When overvoltage, overcurrent, or overtemperature conditions are detected, the fuel gauge IC will command the AFE to turn off the charge and discharge MOSFETs Q1 and Q2. When cell undervoltage is detected, it will command the AFE to turn off the discharge MOSFET Q2 while keeping the charge MOSFET on so that battery charging is allowed.

The main task of the AFE is overload and short-circuit detection and protection of the charge and discharge MOSFETs, cells, and any other inline components from excessive current conditions. The overload detection is used to detect excessive overcurrents in the battery discharge direction, while the short-circuit detection is used to detect excessive current in either the charge or discharge direction. The AFE threshold and delay time of overload and short-circuit detection can be programmed through the fuel gauge data-flash settings. When an overload or short circuit is detected and a programmed delay time has expired, both charge and discharge MOSFETs Q1 and Q2 are turned off and the details of the condition are reported in the status register of the AFE so that the gas gauge can read and investigate the causes of the failure.

The AFE serves an important role for Li-ion battery pack gas gauge chipset solutions. The AFE provides all of the high-voltage interface needs and the hardware current protection features. It offers an I²C compatible interface to allow the gas gauge to have access to the AFE registers and to configure the AFE's protection features. The AFE also integrates cell-balancing control. In many situations, the state of charge (SOC) of the individual cells may differ from each other in a multicell battery pack, causing voltage differences among cells and cell imbalances. The AFE incorporates a bypass path for each cell. These bypass paths can be utilized to reduce the charging current into any cell and, thus, allow for an opportunity to balance the SOC of the cells during charging. Because the IT gas gauges can determine the chemical SOC of each cell, the correct decision can be made when cell balancing is needed.

6.3 Battery Pack ESD Design Considerations

6.3.1 ESD Fundamentals

Virtually all ICs have an internal substrate diode from each pin of the ICs to the ground. This diode is part of the ESD protection structure in the device. A typical device will be protected to 1.5 to 2 kV by this internal structure with human-body-model (HBM). However, typical end-equipment specifications will have a system ESD requirement of ±15 kV of air discharge, requiring additional ESD protection components. Table 6.1 shows the EN 61000-4-2 electrostatic discharge test for both air discharge up to ±15 kV and contact discharge up to ±8 kV. It shows that temporary performance degradation and recovery by operators are acceptable as long as there is no hardware damage failure.

Before we discuss how to prevent device damage from ESD hits, let's look at the following interesting example. In the June 23, 2005, edition of *EDN* magazine, Howard Johnson wrote an article titled "Watery Grave." In the article, he presented the scenario of a person on a lake in an aluminum canoe as a terrible thunderstorm was approaching. Given the following three choices, the reader was asked to select the one that would afford the best chance for survival, assuming there would not be a direct lightning strike, which would be fatal in any case:

- Choice 1: Stay in the canoe.
- Choice 2: Swim to shore.
- Choice 3: Invert the canoe and dive under it for protection, as it becomes a Faraday shield.

A direct hit would be deadly in any case—just as a direct 15-kV ESD strike could be deadly to an IC. When the lightning hits, megajoules of energy pass through the water. If your body is in the water, some of the energy will pass through you, and it only takes a tiny amount of the total energy to kill

Table 6.1
EN 61000-4-2 Electrostatic Discharge Test

Severity Level	Air Discharge	Pass/Fail Acceptance	Contact Discharge	Pass/Fail Acceptance
1	±2kV	Normal operation No degradation; No failure	±2kV	Normal operation No degradation; No failure
2	±4kV	Normal operation No degradation; No failure	±4kV	Performance degradation allowed No data lost. Self-recoverable. No hardware failures.
3	±8kV	Performance degradation allowed No data lost. Self-recoverable. No hardware failures.	±6kV	Performance degradation allowed No data lost. Self-recoverable. No hardware failures.
4	±12kV	Performance degradation allowed No data lost. Self-recoverable. No hardware failures.		
5	±15kV	Temporary performance degration. Recovery by operator is acceptable. No hardware failure.	±8kV	Temporary performance degration Recovery by operator is acceptable No hardware failure.

you. Therefore, the correct answer, of course, is to stay in the canoe because the hull of the boat would divert the current around the person. The same strategy could be applied to protect ICs inside a battery pack from the miniature lightning of an ESD hit. If the pack could be fitted with a metal case, the solution would be clear. Although the solution with the standard plastic case is not quite so obvious, the method is still the same—the current must be diverted around the unit to be protected.

6.3.2   Where Does the Current Flow During ESD Hits?

Most battery pack requirements include surviving multiple ESD hits from both direct connection and air-gap spark discharges. The equipment must generally withstand both positive and negative discharges of at least 15 kV to all connector pins like Pack+, SMD, SMC, and Pack– pins as well as to the case of the battery pack. Most requirements go further than just requiring survival, insisting that there be no observable disruption in performance. Because component ratings are generally much lower than 15 kV, the electronics must include protection components and design countermeasures to reduce ESD damage and upset potential. ESD damage control is generally provided by shunt Zener diodes, transient suppressors, and bypass capacitors. Series resistors may be used to limit the peak current flow. It is noteworthy that 15 kV may arc across the body of some small resistors, reducing or eliminating their effectiveness for limiting current flow. The key to designing circuitry that meets ESD requirements is an understanding of where the peak current flows from the ESD event.

An ESD event usually results in a very fast-rising voltage and current pulse on the line that receives the discharge. During an electrostatic discharge from a human body onto the battery pack connector, the current from the charged source tends to flow into the lowest impedance, like the largest available capacitance, which is that of the cells themselves with respect to earth ground. Naturally, most of the current tends to take the path with the lowest impedance. Wide copper traces, with their low resistance and inductance, become the diverters—able to protect the sensitive electronics from grave danger.

Figures 6.2 and 6.3 show the preferred diverting path for a zap to Pack+ and Pack– [1]. The Li-ion cells, protection MOSFETs, sense resistor, and the pack connector surround the battery management unit (BMU). The single RC filter to the left of the BMU represents one of several connections, and it monitors the voltage of each cell. The connections below the BMU represent various connections from the electronics to the common ground point, which is usually located on the Pack– side of the sense resistor. The resistors and Zener diode to the right of the BMU represent the typical protection network for one of the communication lines. The capacitor in parallel with the protection MOSFET is used to divert the current from flowing to the MOSFET. The dotted line shows the current flow during the ESD hit. In both cases, the preferred path is similar. The copper to the FETS is wide, but then what? The capacitor (usually two in series in the event that one shorts) across the FETS helps to protect them. However, this can only be realized if the copper traces to the capacitor

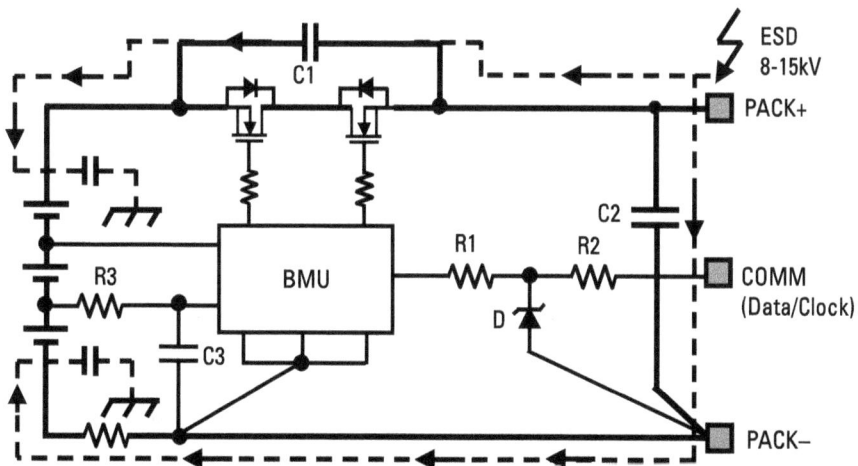

Figure 6.2   ESD hit to PACK+ connector.

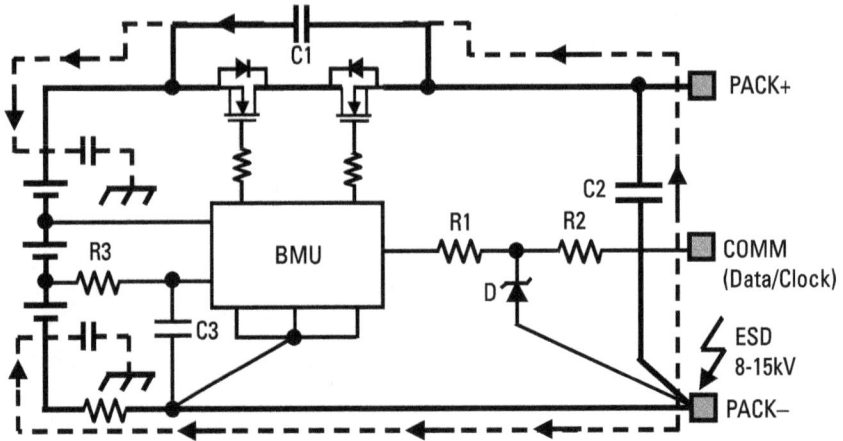

Figure 6.3   ESD hit to PACK– connector.

are also wide enough to offer the required low resistance and inductance. The capacitor, which usually has two in series, between Pack+ and Pack–, is equally critical. It is desirable that current from a hit to Pack+ be diverted, as much as possible, away from the MOSFETs and their associated nodes, which lead into the electronics. The copper between the pack connections and the capacitors must be short and thick.

Figure 6.4 shows the current path when the ESD hits the communication lines. Note that the desired current path is through the first series resistor, which is used to limit the current, through the capacitance of the Zener diode, then on to the wide Pack– copper trace. Keeping the Zener close to the pack connector and using sufficient copper width ensure that the BMU is protected. In the case of a negative polarity zap, current flows out of the BMU in parallel with current through the Zener. The resistor on the BMU side limits the ESD current to a safe level.

Whereas wide traces help lower the inductance of copper traces, they still appear quite inductive at ESD pulse frequencies. The ESD event may cause a 1-ns rise-time voltage pulse of several thousand volts and/or more than 30A of momentary current flow. The fast-rising voltage spike can capacitively couple into any trace and components adjacent to the affected line. The fast-rising current flow from the discharge creates a large inductive voltage drop along the path of the current flow. It also creates a magnetic field from the current flow that can induce transients into other circuitry through the nearby components and PCB traces. In addition, at high frequencies, the diverting path can act as the primary of a current transformer, injecting unwanted and potentially disruptive currents into adjacent copper loops that feed into sensitive electronics such as the ultra-low-power microprocessors used in fuel gauge systems.

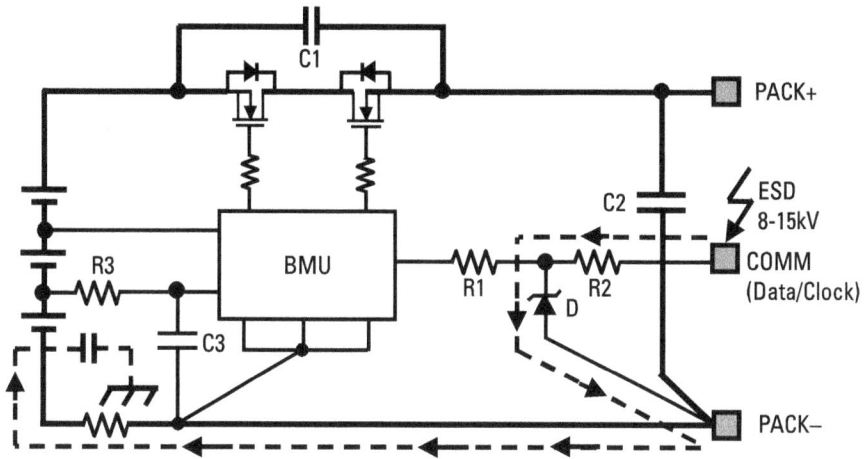

Figure 6.4 ESD hit to the Data or Clock pin (communication line).

The best defense against this sort of assault is to physically separate the high-current path from the sensitive electronics. Although this may not be feasible in many battery pack designs, it is an ideal goal for rugged design. High-current inrush pulses and ESD pulses do not mix well with ultralow-power electronics. Both inductive and capacitive coupling must be considered in the layout.

6.3.3 ESD Design Hardening

Start the hardening process at the connector. One popular technique for improving ESD susceptibility is to build a spark-gap structure in the outside PCB layer behind the battery pack connector [2]. Figure 6.5 shows the recommended spark-gap pattern for a battery pack connector. This is a low-inductance (wide) ground PCB that runs close to the PCB pad connections for the other connector pins. The PCB structure provides a small clearance between points, or corners, in the PCB to encourage a breakdown from a high voltage across the clearance. The clearance must be kept free from the solder mask, because the solder mask would increase the voltage breakdown of the gap enormously. A 10-mil gap has a breakdown voltage of about 1,500V. This breakdown voltage is typically less than the damage threshold of IC inputs. The spark gap tends to clamp the peak voltage on the connector pins other than ground and divert much of the charge to the ground PCB. If the ground PCB is handled carefully, this approach can significantly improve the ESD susceptibility of a design.

Keep the PCB connected to the top and bottom of the cell stack away from all sensitive components. If the connections to the cell stack run the full length of the PCB, an upset from an ESD event is much more likely to occur

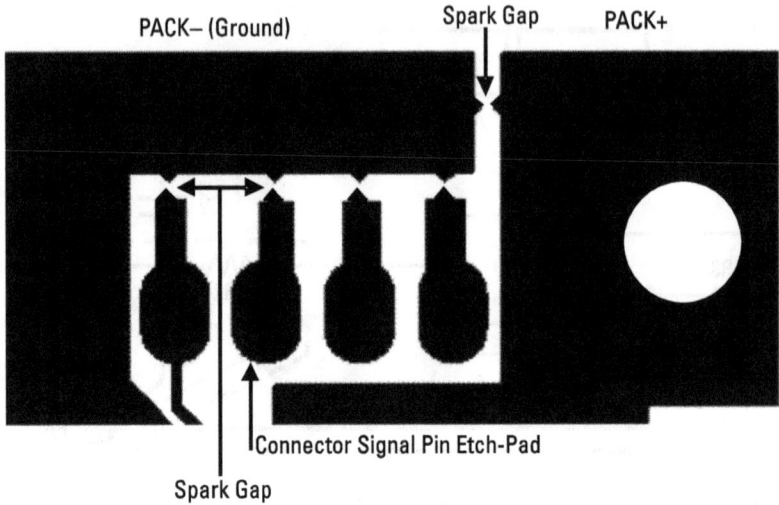

Figure 6.5   Recommended spark-gap pattern for a battery pack connector.

due to the capacitive and magnetic coupling to nearby components and PCB traces. If the Pack+ and Pack– connections can route through the protector FETs and sense resistor and then immediately leave the PCB and connect to the cell stack, a large portion of the PCB may be relatively free from the coupling resulting from an ESD event.

The recommended design practice is to separate the high-current ground PCB from the low-current ground PCB. Even a small inductance on the high-current ground PCB will develop a large potential across the length of the connection due to the extremely fast di/dt from the ESD event. If sensitive circuitry has connections to the ground at different points along the high-current ground path, there may be a large differential voltage between these connections during an ESD event. This differential voltage may allow some inputs to be momentarily pulled lower than ground, and the resulting substrate current flow can upset the circuit performance. The best way to handle this issue is to connect all of the low-current grounds together and then tie the low-current ground to the high-current ground at a single point. Make sure that none of the ESD protection components, such as shunt Zener diodes or transient suppressors, tie to the low-current ground; instead they should tie to the high-current ground, as shown in Figure 6.2. This greatly reduces the possibility for an ESD event to cause current flow through the low-current ground and to create differential voltages among different ground connections.

Keep bypass capacitor leads short. It is very important to avoid canceling the high-frequency capability of a good ceramic capacitor by adding a trace inductor in series with it. This is what happens when care is not taken to keep both connections to the bypass capacitor short and wide. A long connection on the ground side of the capacitor is just as bad as a long connection on the signal side. Use of a ground plane, where practical, makes this a lot easier. If an electrolytic capacitor is required to obtain the needed bulk capacitance, add an additional ceramic capacitor in parallel to take care of the high-frequency by-pass requirement. The frequency components in an ESD pulse are so high that even ceramic capacitors may seem inductive. Some critical circuits may benefit from a small-value (68- to 100-pF) ceramic capacitor in parallel with a larger (0.1-μF) bypass ceramic capacitor because the impedance of the small-value capacitor is usually much less at the higher frequency due to much lower series inductance. Long connections to series components may not be a problem and, in fact, may add some inductance that will aid in decoupling. For example, some designs use small inductors in series with VCC connections to aid in decoupling an ESD transient from a sensitive circuit.

Placing a 0.1-μF ceramic capacitor across the Pack+ and Pack– connector pins very close to the connector and using short and wide PCB traces can reduce the ESD susceptibility. This placement will provide an alternate path for a portion of the current pulse from a Pack+ or Pack– ESD hit and reduce the peak current amplitude through the PCB traces to the battery cells. This may provide some improvement in the peak ESD voltage transient level that the pack can withstand.

Protecting communication or other interface signals such as Clock and Data lines with Zener diodes and/or transient suppressors is mandatory. The shunt suppression component should tie to the high-current ground as mentioned previously. It is very important to ensure that the impedance of the path intended to shunt the transient to ground be significantly lower than the series impedance to the device input being protected. A long, thin connection to the transient suppressor may severely reduce the effectiveness of the protection components. It is generally helpful to add some series impedance both before and after the shunt Zener diode or transient suppressor. Resistance between the pack connector and a shunt Zener diode could reduce the peak current through the latter and keep it from failing. The Zener diode normally fails as a short and will render the interface signal useless. The series resistance may need to be a larger package style that will withstand a larger voltage without arcing. This resistance may also protect the Zener diode from failing due to a momentary short from the interface signal to Pack+. The larger resistor package will also withstand a momentary overload for a longer time. Resistance between a shunt Zener diode and the IC input is also effective to reduce the potential for upset. The IC input will generally have an ESD protection structure that has a diode to VSS. If a negative ESD transient is applied to the input, the shunt Zener diode and ESD protection diode in the IC will share the current. A little resistance between the Zener diode and IC input will force most transient current to flow through the Zener diode instead of sharing it with the IC. Current flow through the substrate of the IC is very likely to cause an upset condition in the IC, so removing that possibility is a big help.

Another design approach to enhance performance is to implement a reset strategy that does not always upset the information displayed to the user. If critical data are maintained even with a reset, an upset due to an ESD event may not cause any significant disruption to the user. This strategy may employ redundant copies of critical information or checkbyte values that may be used to determine if the critical data are still useful after a device reset.

6.3.4   Pack Insertion Issues

Some designs have communication and/or interface signal lines to circuitry in the battery pack. In many cases, the ESD protection on these lines does not clamp a positive transient to less than the VCC of the battery electronics. If the circuitry in the battery pack contains a substrate diode from the communication line to VCC as shown in Figure 6.6, it is possible to disrupt the VCC supply when plugging in the battery pack. This disruption may cause improper operation of the battery pack electronics. If the host system is not applying a charging potential to the host-side pack connector, the capacitance across the battery connections will be discharged before the battery pack is inserted. Most battery connectors do not have any provision for ensuring that the Pack– or ground connector pin mates first. If the Pack– pin connects last when the battery pack is plugged in, there is a path to pull up the VCC in the battery electronics temporarily until it almost reaches the Pack+ terminal potential. The electrical path to pull up the battery pack VCC passes through the

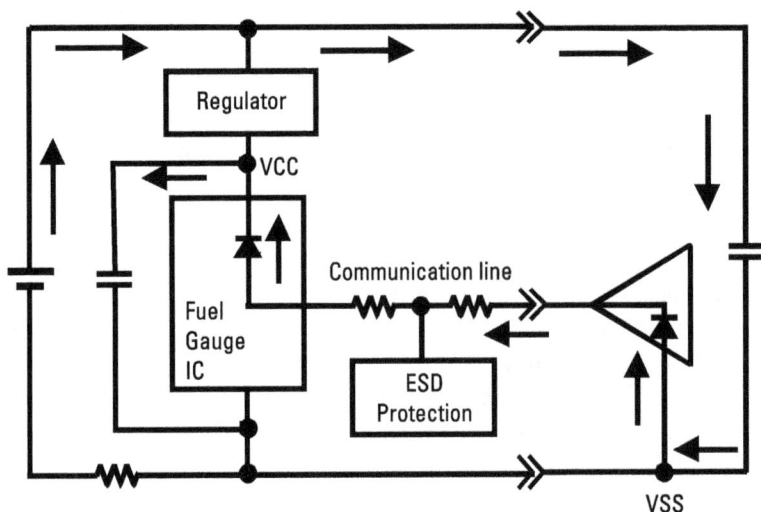

Figure 6.6   Pack insertion issue.

host capacitance from Pack+ to Pack–, through a substrate diode in the host interface driver from VSS to the communication or interface line, and through a substrate diode from this line to VCC in the battery pack circuitry. The best design practice is to use circuitry in the battery pack that does not have an internal substrate diode to VCC. This has a side benefit of preventing the battery pack electronics from being able to clamp the communication line to a low VCC value resulting from a depleted battery. The ESD protection circuitry on the communication line should also clamp the voltage on the line to less than the maximum allowable voltage. This will limit the peak transient voltage and prevent damage to the battery electronics if the ground pin makes the last connection during pack insertion.

From this discussion you can see that battery circuit design and layout are considerably more critical than might be expected. The combination of battery requirements includes high-amplitude ESD to connector pins and exposed surfaces, coupling from an ESD event to nearby PCB traces and components, heavy load currents, plugging and unplugging with power on the connector pins, multiple circuit ground references including high-current grounds, measurement of very small signals, and thermal management. Meeting these requirements and implementing the design on a circuit board that may be almost too small to hold the required components provides an extremely difficult challenge. The key to a successful design is the recognition of the various issues prior to starting the design and close control of the PCB layout by the design engineer. A good design involves a layered approach; removing any one layer will reduce the effectiveness of the others.

6.4 Electromagnetic Interference (EMI) Solutions

6.4.1 EMI Solutions in the Battery Management Unit

While an ESD event can be characterized as a high-power microwave pulse, other sources of radio-frequency (RF) energy can also have a negative effect on an unprotected battery pack. RF energy can be transported into a battery pack by either radiation or conduction. The cells and their leads can act as an antenna, or copper traces on the PCB itself can be the receiving antenna. Antennas are most effective at multiples of a signal's wavelength. A cell phone operating at 1,800 MHz has a fundamental wavelength of 16.7 cm. A nice half-wave antenna is 8.3 cm, while an effective quarter-wave antenna is only 4.2 cm. For this reason, RF testing of a new battery pack design is highly recommended to ensure its dependability in common RF environments such as those created by cell phones and other two-way radios. Procedures and parameters for EMI testing are given in [1]. The field strength should be 30 V/m.

Rectified RF can cause a number of problems, including errors in measuring voltage, temperature, and current in the BMU. Microcontroller misoperation and unintended fuse blowing are also possible. If any of these effects are observed during testing, it may be relatively easy to bypass the receiving semiconductor input with one or more small RF capacitors. The 22- to 100-pF range RF capacitors have a very low impedance at very high frequency (VHF) and ultra-high-frequency (UHF) radio-frequencies for bypassing the RF frequency signal.

In the fuse-blowing SAFE circuit shown in Figure 6.7, the use of RF capacitor C1 from gate to ground is generally good practice, especially for RF immunity, but may be omitted if desired since the chemical fuse is a comparatively slow device and will not be affected by any submicrosecond glitches that may come from the SAFE output during the cell connection process. R3 is used as a damping resistance for suppressing the RF signal, which prevents the false turn-on of MOSFET Q1 with R1 and R2 as a discharge path. It is important to observe that this network may rectify strong RF signals and produce an unwanted positive DC level on the gate of the FET Q1 in some layouts. A real example is that turning on a 2W walkie-talkie next to a circuit board can turn on Q1, falsely causing the fuse to be blown. It turns out that the unshielded circuit trace from the IC SAFE pin to the anode of D1 can conveniently pick up RF power and slowly increase Q1's gate voltage until it turns on. Another possible solution is to ensure that the trace on the anode side of D1 to the SAFE pin is kept short and/or is shielded by ground on both sides.

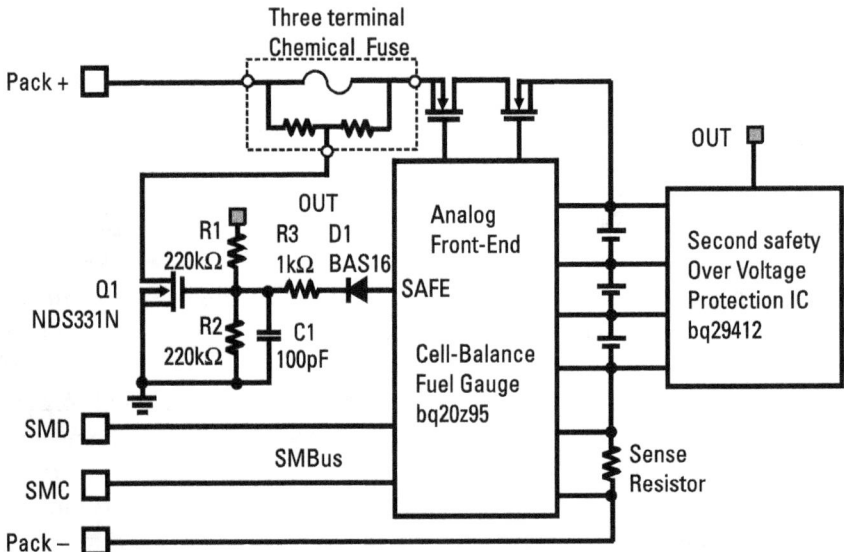

Figure 6.7   Fuse and SAFE protection circuit.

6.4.2 EMI Design Considerations in Battery Charging System Applications

The threat of generating EMI from the high frequencies inherent to fast-changing voltage and current waveforms has always been a serious concern, especially in switch-mode power supplies. To minimize total solution size and improve the power conversion efficiency in a battery-powered system, switch-mode battery chargers and DC-DC converters are very popular. The passive components such as inductors and capacitors take up about 70% to 80% of the overall board area in today's practical switch-mode power converter designs. These passive component sizes are tightly related to the switching frequency. For the same percentage of inductor ripple current, the inductance is inversely proportional to the switching frequency. The higher the switching frequency, the smaller the inductance and capacitance. Operating the switching frequency at 1 to 3 MHz is very common compared with the few hundred kilohertz used about 10 years ago, due to continuous improvement of the ultrafast switching MOSFETs and new magnetic core material improvements. The higher switching frequency allows all subsystems to be placed in proximity to the systems. Therefore, designing for electromagnetic compatibility (EMC) has become a requirement, which could be just as important as meeting a given set of power conversion performance specifications. The purpose of this discussion is to address some of the more important principles and techniques involved.

First, note that although we have used the terms EMI and EMC interchangeably, it should be clear that one is the inverse of the other. The definition of EMI is "the interference of one piece of electronic equipment on the operation of another by means of electromagnetic energy transfer." This definition of EMI involves three elements:

- A generator or source of electromagnetic energy.
- A coupling method (magnetic or electric field) of that energy between equipment.
- A receptor circuit or a victim circuit whose operation is impacted by the transmitted energy.

All three elements must be present for EMI to take place at the same time. There will be no EMI if either of them is removed. Shielding or separation may solve a specific interference problem by attacking or eliminating the coupling or susceptibility part of the system. However, the preferable approach is to remove or minimize the EMI source.

EMI is often neglected in electronic circuit design education and looks quite complicated in practice, and it is usually dealt at the end of circuit design with trial and error. Dealing with EMI is a very time-consuming, challenging process for many circuit designers. The basic principles for both causing and solving EMI are relatively simple. Fundamentally, we need to recognize that the fields are caused by rapidly changing currents and voltages. While these characteristics are quantitatively described by Maxwell's equations, we need only to know that electronic noise may be induced by coupling between circuit elements through the action of either a magnetic or an electric field.

A fast-changing current, which creates a magnetic field in one loop, induces a voltage in another loop according to:

$$u = M \frac{di}{dt}$$

where M is the mutual inductance between the source and the victim.

Similarly, a changing voltage, which creates an electric field, induces a changing current in another conductor according to:

$$i = C \frac{dv}{dt}$$

where C is the coupling capacitance between the source to the victim.

These equations show that whenever there is rapidly changing current di/dt—as in the conductors in series with power switching devices—we can expect to see an induced voltage across other conductors coupled by a mutual inductance. Whenever there is a high changing voltage dv/dt—as on the drain of the power switching MOSFETs—we can expect to see current induced into another path through any coupled parasitic capacitance. From the above two equations, we can see that the tighter the coupling between changing current or voltage source and victim, the stronger the generated EMI. In addition, the faster the changing current or voltage, the worse the EMI that is created. One of the most obvious ways to reduce generated EMI is to slow down the switching speed or transition, but this results in increased switching losses. Therefore, a design trade-off is required between minimizing the EMI through slowing down the switching transition and compromising the power conversion efficiency.

6.4.3 Measuring the EMI

When we discuss a noise-generating system, the transmission of the noise out of the system is by either or both of two separate coupling paths. The EMI noise could be either the radiated energy from the system itself through magnetic or electric fields or the conducted energy flowing in either the input, output, or control lines. These conducting and radiating EMI noise sources are considered as separate and distinct and are usually specified separately when evaluating the external characteristics of a definable system. One helpful characteristic, however, is that the efficiency of noise coupling is frequency dependent—the higher the frequency, the greater the potential for radiated EMI. In contrast, at lower frequencies, problems are more likely to be caused by conducted noise. There is universal agreement that 30 MHz is the crossover frequency between conducted and radiated EMI. Most regulating agencies throughout the world have established measurement standards that specify the evaluation of conducted EMI at frequencies up to 30 MHz, with a separate set of standards applicable above 30 MHz for radiated EMI.

Conducted noise is primarily driven by current, but is measured as a noise voltage with a 50-Ω current shunt resistor. The Federal Communications Commission (FCC) cares only about the AC input power lines because the noise currents could be easily coupled to other systems through the power distribution network. The maximum frequency of interest for conducted EMI is 30 MHz, and the minimum frequency limit can vary. In the United States and Canada, 450 kHz is the minimum frequency. However, many international specifications specify a lower frequency limit of 150 kHz. Some telecom specifications require testing the EM starting at 10 kHz.

The radiated noise specifications start with a lower frequency limit of 30 MHz and an upper frequency limit of 1 GHz or higher. The conducted noise can be evaluated with a spectrum analyzer, while the radiated noise requires the measurement of magnetic or electric fields in free space, causing the testing to become much more complicated. For this reason, radiation testing is usually performed by a specialized test facility, where variables inherent in the testing environment can be very closely controlled.

System usage defines "classes" where Class A designates industrial and commercial applications, and Class B includes residential usage. Class B limits are more stringent because systems made for the home are likely to be closer to each other. Residential users usually have fewer options available for dealing with EMI issues.

In the United States, the FCC is responsible for EMI control through the Code of the Federal Register (CFR), Title 47. In this document, Part 15 controls information technology equipment (ITE), Part 18 covers industrial, scientific and medical equipment (ISM), and Part 68 regulates equipment connected to a telephone network.

International EMI specifications have been led by the International Electrotechnical Commission (IEC), which has published a European Union generic standard for emissions (EN50081), and the French-led Comité International Spécial des Perturbations Radioélectriques (CISPR). This latter body has issued an EMI specification, CISPR Publication 22, that has gained worldwide acceptance. In addition to some limit value differences, the differences between the FCC and CISPR specifications include a lower frequency range for CISPR conducted noise measurements, and radiation tests made at a fixed distance of 10m versus 3 to 30m for FCC testing. By extrapolating the FCC radiation limits to 10m (linearly proportional to 1/distance). The comparisons can be better illustrated with the frequency plots for both conducted EMI and radiated EMI, as shown in Figures 6.8 and 6.9, respectively.

Note that the units in these specifications are given as dBµV for direct measurements of conducted noise and dBµV/m for field strength measurements of radiated noise as sensed with an antenna. These are decibel values giving the ratio between the actual voltage measurement and 1 µV. The calculation is given by

$$dB\mu V = 20 \log_{10}\left(\frac{V}{1\mu V}\right)$$

and it applies to both volts and volts/m.

6.4.4   Conducted EMI

To study conducted EMI noise on a power line, we need to separate the high-frequency noise signals from the input current. The device used to separate such high-frequency noise is called a line impedance stabilization network (LISN). Its schematic is shown in Figure 6.10.

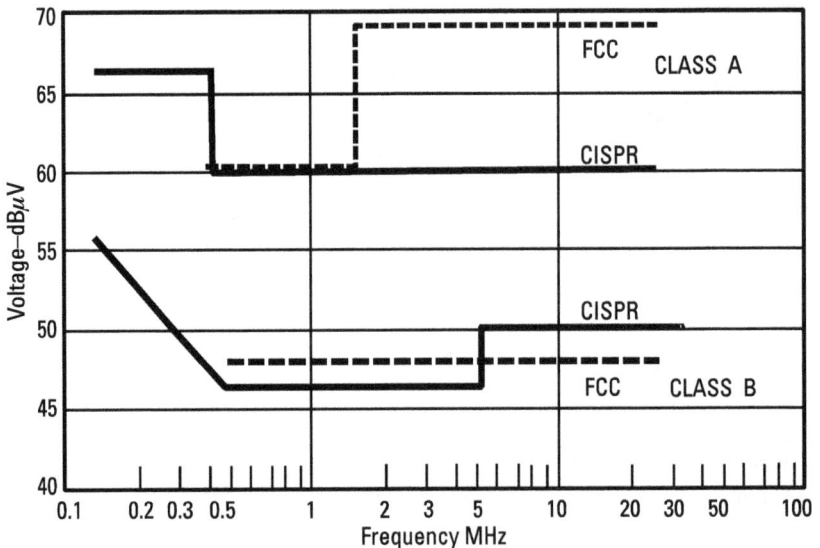

Figure 6.8   Conducted EMI noise limits (FCC Part 15 versus CISPR 22).

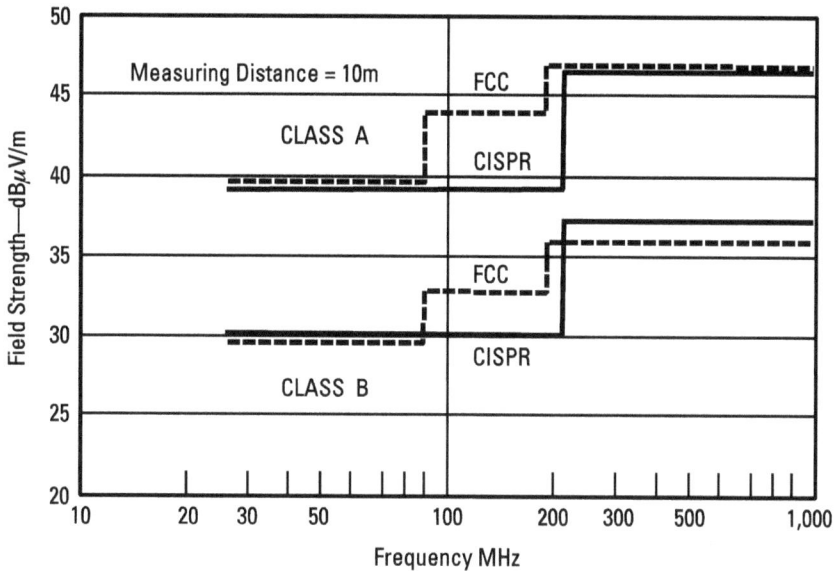

Figure 6.9 Radiated EMI noise limits (FCC Part 15 versus CISPR 22).

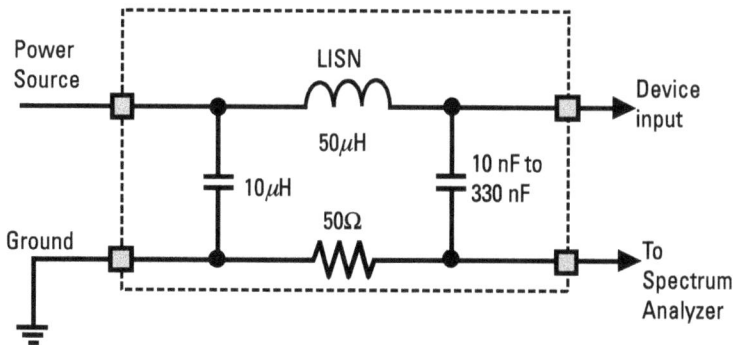

Figure 6.10 LISN circuit for measuring conducted EMI.

All measurements are made with respect to ground. The spectrum analyzer is used to measure the noise current through a 50-Ω resistance. A LISN network is added to each of the input power lines and the noise signals are measured with respect to ground.

Conducted noise consists of differential mode (DM) noise and common mode (CM) noise. The DM noise is measured between the power feed and its return path, whereas CM noise is measured between each of the power lines and earth ground. These two EMI noises are inherent from the basic operation of a switching power converter or switch-mode battery charger. The on and off switching behavior of the power switches causes a fast current change of di/dt

in the differential current at both the inputs and outputs of the power supply. Of course, input and output filters would eliminate any high-frequency noise external to the power supply, but neither can do the job completely. So residual ripple and switching spikes exist as a differential mode noise source with current bidirectionally flowing into one terminal and out the other.

There are also sources of rapidly changing voltage in the switch-mode battery charger, which can couple the noise through parasitic capacitance to earth ground. In the buck-based switch-mode charger, the drain of the power MOSFET typically has a very fast voltage change of dv/dt as shown in Figure 6.11. This type of noise in the ground path is considered to be CM noise and is measured with respect to earth ground.

The DM and CM noise currents have two different paths as shown in Figure 6.12. There are two LISN devices in series with both the power line input and its return. Both CM and DM modes of current are present in each LISN. The power line LISN measures CM + DM, while the LISN on the neutral return measures CM – DM. There are circuit networks used to separate CM and DM signals [3], but the specifications typically do not differentiate. The total noise at each input has to be measured because, with the possibility of multiple current paths, there is no reason to assume that the values of the CM and DM contributions at the two inputs are identical.

6.4.5 Approach for Minimizing Conducted Differential Noise

A filter is usually used to minimize the conducted differential noise. This is easier said than done, however. Filter performance will be analyzed in terms of voltage attenuation between the input terminal connected to the power source, and the filter output connected to the switch-mode battery charger circuit input as an example.

The first point to keep in mind is that we are shooting for minimizing the conducted differential mode noise. The filter has to connect across the

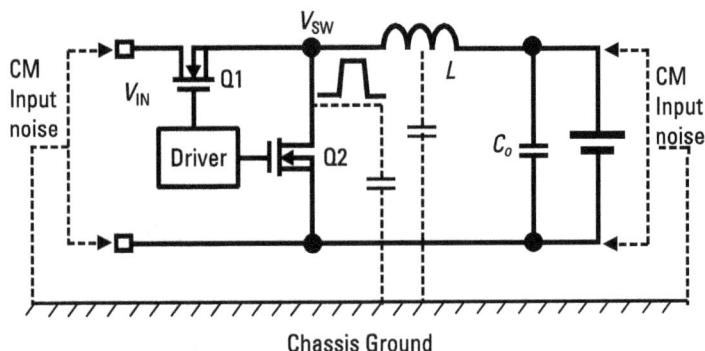

Figure 6.11 CM noise from drain of synchronous MOSFET to earth ground.

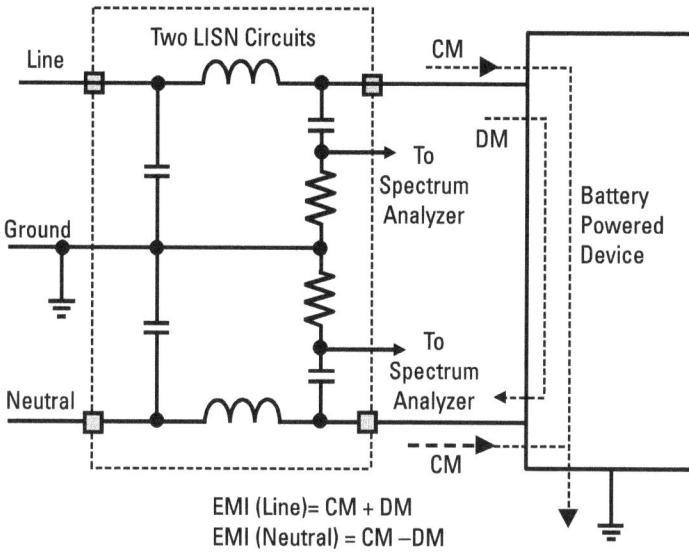

Figure 6.12 DM and CM current flows.

differential lines. Figure 6.13(a) shows an ideal LC low-pass filter at the input, where V_{IN} is the power source and V_{OUT} is the power input connection to the battery charger; the capacitor negative terminal has to connect to the power return line, not the earth ground! In reality, there is no such ideal filter that could yield the perfect attenuation curve shown in Figure 6.14. The actual filter includes the parasitic capacitance in parallel with the inductor, and parasitic equivalent series inductance (ESL) and equivalent series resistance (ESR) in series with the capacitor shown in Figure 6.13(b). If these parasitic components are included, the filtering attenuation is not as good as the ideal filter. The ESR of the capacitor introduces an ESR zero, and the combination of the ESR of the capacitor and inductor introduces another zero. In addition to this, the inductor actually becomes a capacitor due to its parasitic capacitor effect, which makes the filter less effective. To continue to further minimize the parasitic effects, it is necessary to look for a component with a smaller parasitic effect. Typically, the smaller the output capacitance, the smaller its parasitic effects. Therefore, the common practice is to use a few output capacitors in parallel to achieve good attenuation.

6.4.6 Approach for Minimizing Common Mode EMI Noise

As we discussed earlier, the CM noise is mainly generated by parasitic capacitance from a high-voltage switching node to the earth ground. A simplified example of one of the major EMI problem sources is shown in Figure 6.15. A switch-mode synchronous buck-based charger was used as an example with a

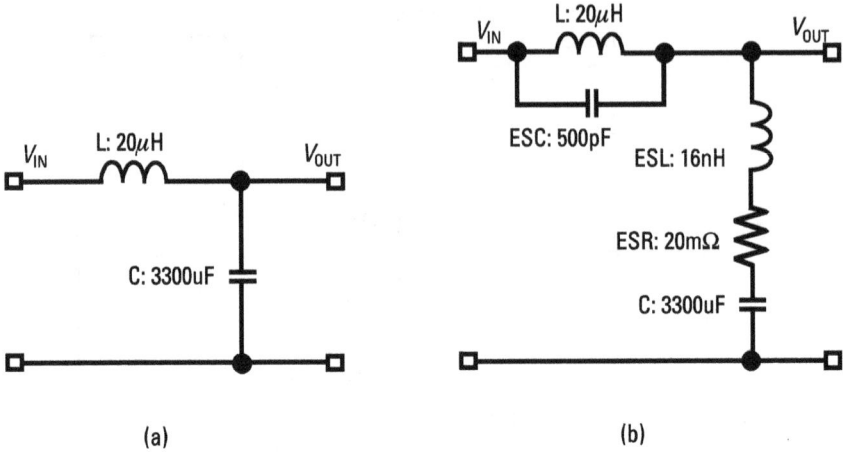

Figure 6.13   Differential EMI filter.

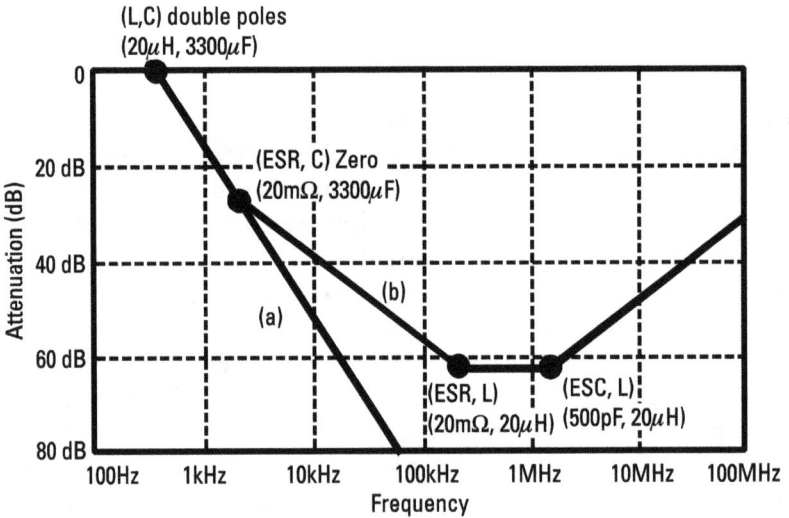

Figure 6.14   Differential EMI filter attenuation.

600-kHz switching frequency and 20V input voltage. The top control MOS-FET has an approximately 20-ns turn-on time, during which the switching node voltage gradually increases from 0 to 20V. In the meantime, such a voltage increase will charge the parasitic capacitance from the drain to earth ground resulting in the CM noise. The parasitic capacitance depends on the layout and heat sink used, and it is typically about 10 to 40 pF. Assuming it has 20-pF parasitic capacitance, the $C dv/dt$ creates a peak current of ±20 mA.

Figure 6.16 shows how this injected CM noise current completes its path back to the input through the two 50-Ω LISN resistors in parallel, thereby

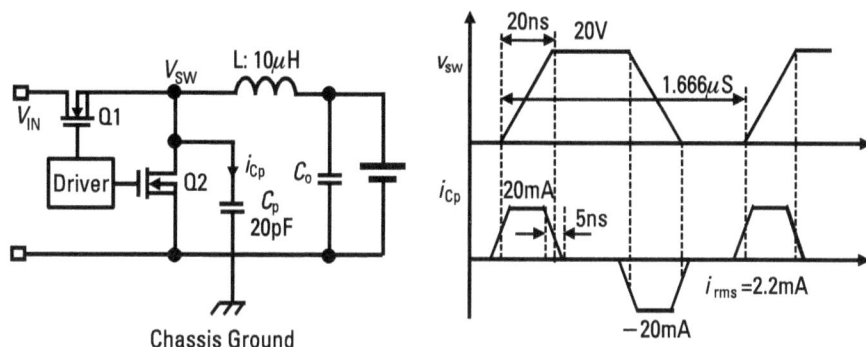

Figure 6.15   CM noise with capacitive coupling from switching node to ground.

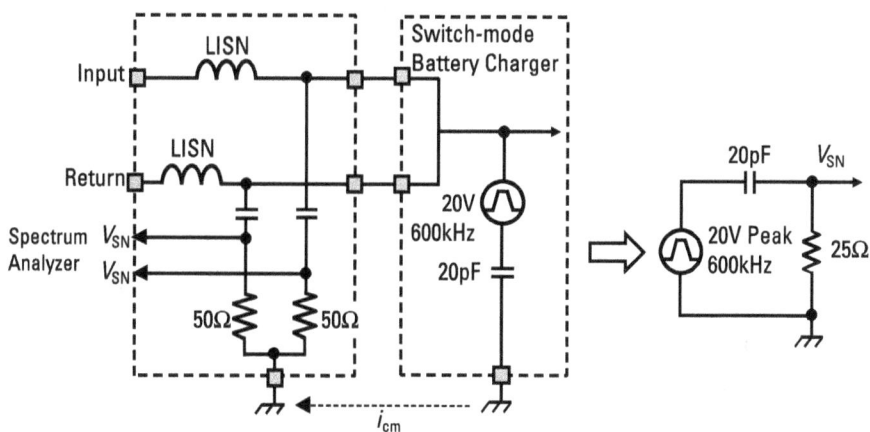

Figure 6.16   CM noise loop with LISN and its equivalent circuit.

creating a noise voltage at each LISN output. Assume that the buck-based char-ger has 50% duty cycle. Based on the equivalent circuit, the V_{SN} has a CM noise voltage of 26.4 mV across the parallel LISN resistors. Though that may seem like a small amount, it fails to meet the FCC noise limit of 1.0 mV (60 dBμV) for Class A products at 600 kHz; the limit for Class B is even lower at 0.25 mV.

To reduce 26.4 mV to less than 1.0 mV, we need to insert a filter that generates an attenuation of 28.5 dB at 600 kHz. One way of achieving this goal is to add a series common mode inductor. Working backward, we can calculate the required inductance from the reactive impedance at 600 kHz as shown in Figure 6.17. The required inductance would be 93 mH; we cannot have more than 0.75-pF parasitic capacitance across the inductance. This is unlikely to be achieved.

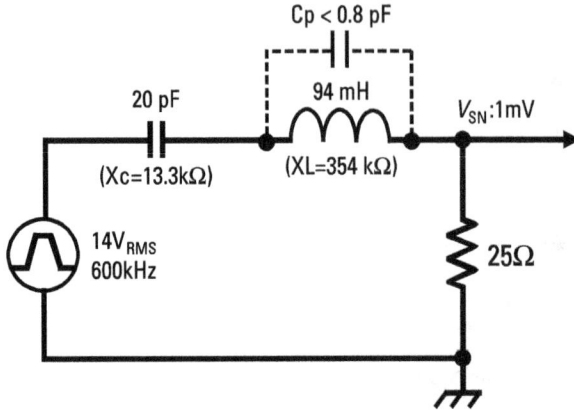

Figure 6.17 Achieving 28.5-dB attenuation by inserting an inductor with an unrealistic parasitic capacitance.

Therefore, we finally come up with the best solution for a CM input filter, which includes both inductance and shunt capacitance, but with reasonable values for each, even considering expected parasitic values like those shown in Figure 6.18. The first attenuation stage is composed of shunt capacitor C1 for achieving a decent attenuation. The second stage is composed of the inductor for achieving additional attenuation. From this design, the parasitic capacitance now required is less than 180 pF, which can be easily achieved. Of course, in this application the shunt capacitors are connected to ground instead of differentially.

Figure 6.19 shows the most commonly used input filter. It combines both DM and CM filters, with C_{c1} and L_c forming the CM filter for minimizing the CM ground noise with the two windings of L_c built onto a single core. To save costs and improve solution size, the leakage inductance of the CM filter is used

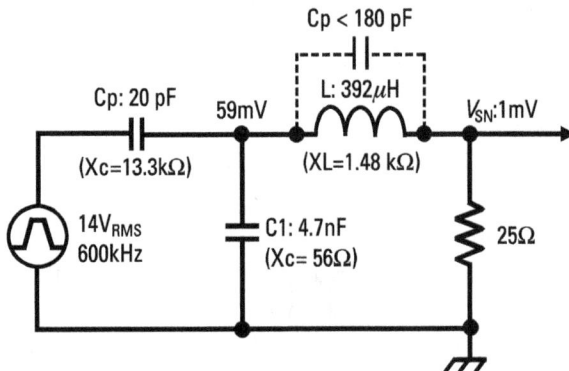

Figure 6.18 CM filter with practical filter inductance.

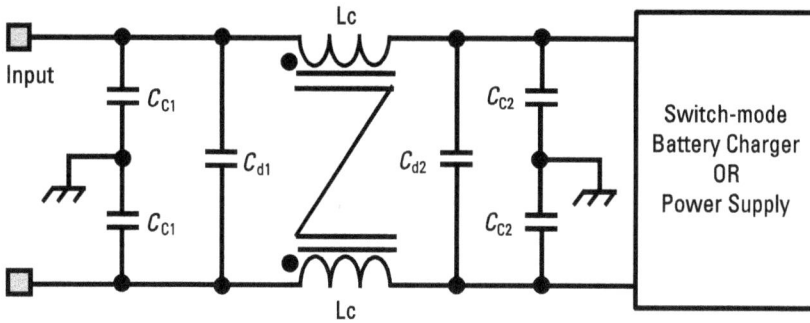

Figure 6.19 Simplified CM and DM filter configuration.

as the differential inductance of the differential filter along with the differential capacitor of C_{d1} and C_{d2}.

6.4.7 Minimizing the Radiated EMI

In principle, each type of noise is separately discussed and treated. In a real electronic system, however, particularly a switch-mode battery charging system or power supply, EMI energy can be transformed back and forth between the conducted and radiated EMI form. If the noise energy is conducted in a wire or PCB trace, or a loop, an electromagnetic field is generated and creates radiated EMI. If there is mutual inductance or capacitive coupling to another conductor, then the radiated energy is transformed back to conducted noise, but in a different location in the system. Any conductor, trace, or closed loop actually acts as an antenna, and an antenna can both send and receive radiated signals.

Testing for radiated electromagnetic compatibility is a much more complicated process than testing for conducted noise. From the EMI specification, radiated EMI frequency ranges from 30 MHz to 1 GHz. All of the test equipment becomes more critical and sensitive to such noise at such high frequency of above 30 MHz than at a low frequency of a few hundred kilohertz. In addition, the test environment has to be well-controlled and typically requires an RF screen room to shield the test setup from any extraneous RF signals from other noise sources. In either case, a knowledgeable operator is crucial in order to obtain reliable data. The type of antenna, its distance and orientation with respect to the device under test, and its ability to sweep all radiating angles are all important parts of the test conditions. A schematic representation of a radiated EMI test setup is shown in Figure 6.20.

To minimize the radiated EMI, the layout is actually one of the most important parameters. The high pulsating current loop acts as an antenna and has to be kept as small as possible. For a buck-based battery charger, Figure 6.21

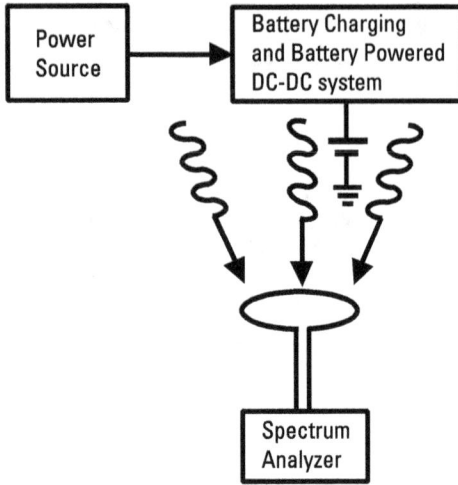

Figure 6.20   Radiated EMI test setup.

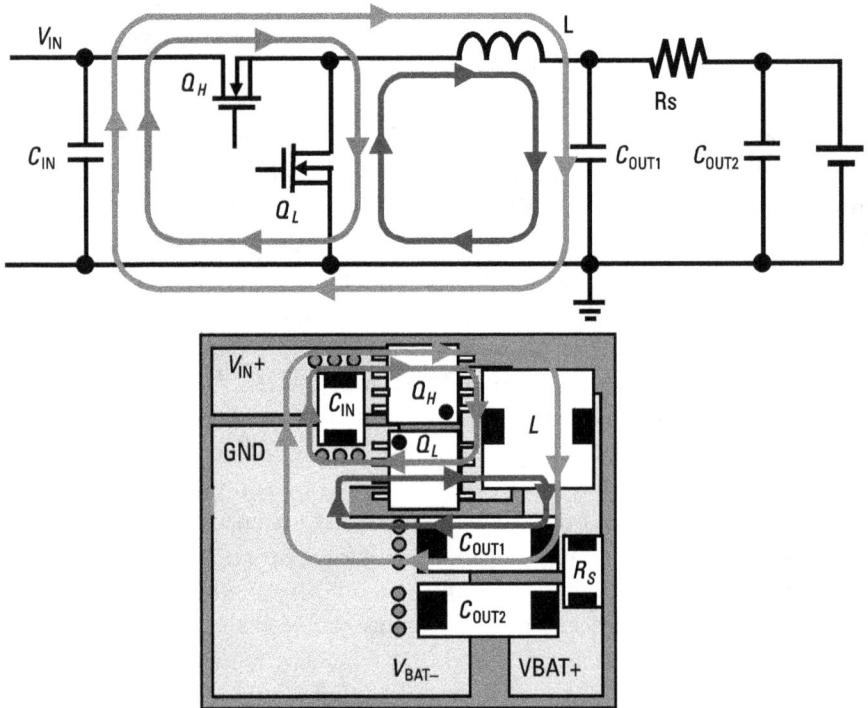

Figure 6.21   Layout example of a buck-based charger.

shows a good layout example for minimizing the loop area. Another important aspect is that the trace of the switching node should be kept as small as possible.

6.5 Power Components and PCB Thermal Design Considerations

With shrinking form factors and increasing functionality and performance, portable battery power equipment is demanding more power than ever before within a limited board space, making thermal management one of the most critical design challenges today. Thermal management can be composed of system power management, battery power thermal management, and traditional thermal design.

Battery management from a thermal perspective is a major concern since excessive high temperature accelerates battery degradation, and may cause thermal runaway and explosions in Li-ion batteries. The switch-mode battery charger typically integrates all power MOSFETs, which creates a big challenge due their power dissipation. The typical semiconductor IC case temperature is usually required to less than 100°C, and sometime needs to be lower than 60°C at ambient temperature of 45°C. One option is to design the battery charging system with a lower on-resistance for the MOSFETs to limit power consumption at the expense of solution cost. Another solution is to actively introduce a thermal regulation loop for regulating the junction temperature of the charger ICs by reducing the charging current with dynamic power management control when the IC junction temperature reaches its thermal regulation threshold. The latter control method is very flexible for achieving desirable thermal performance. The bq24190 charger is a good example of this type of IC.

Thermal management also involves system-level heat-flow management to effectively remove heat from a heat source, to the PCB, and to enclosures through thermally enhanced packages and thermal vias.

The traditional thermal management design remains a challenging task for system designers. Heat generated from the system must be redistributed and dissipated, because portable devices have very limited space and air flow. Due to the size and space constraints of portable devices, a fan and/or traditional heat sink devices cannot be used to transfer heat generated by the power components.

Today, with thermally enhanced surface-mount technology and ICs with power pads, the PCB has become an integral contributor to heat removal. The thermal impedance provided by an IC manufacturer is tested under certain conditions, such as 1-in.² copper and thermal vias under the thermal pad. For example, a 20-pin QFN 4-mm package has a thermal impedance of 32°C/W. However, you may get a much higher thermal impedance if there are no thermal vias and/or not enough PCBs connecting to the thermal pad. Such common design mistakes include inadequate thermal optimization of the PCB layout, lack of thermal vias, and improper placement of components on the PCB and overall enclosure design. All of these mistakes create heat-transfer problems.

PCB design is rapidly becoming part of the critical design stage in thermal management. Copper is a good thermal conductor when used along its length for a large cross section. At least one full copper layer is needed to spread the heat. The larger the perimeter of the power pad (circumference), the greater the cross-sectional thermal conduction area, the lower the PCB's thermal impedance, and the cooler the PCB and ICs.

The top PCB layer usually has many external components connected to the IC pins, so there is little available top PCB area that can be connected to the thermal pad to remove the heat from the power components. One approach for removing heat is to use multiple thermal vias. This removes heat generated by the power components to the bottom PCB layer. Heat transfer through a typical 13-mil thermal via with filled solder provides a thermal impedance of 80°C/W. Ten 13-mil thermal vias under the power IC thermal pad have only 8°C/W thermal impedance. Therefore, multiple thermal vias are strongly recommended to enhance heat extraction from the heat source.

Figure 6.22 shows the layout of a typical surface-mount power device. For example, a high-current battery charger IC with integrated power MOSFETs can dissipate over 1W. In order for the IC or MOSFET to spread the heat over the PCB, enough copper with a number of thermal vias should be connected to the thermal PAD. On a multilayer PCB, if the power ground does not have enough surface area or metal for heat dissipation, then there will not be enough surface or thermal conductivity for the heat to dissipate. As a result, heat may become trapped, causing thermal-related hazards. To combat this, engineers should practice careful board layout. For instance, they should consider enlarging the power ground with more metal, such as two ounces of copper PCB to avoid thermal traps on the PCB.

Now let's look at a design example. The goal is to spread the heat out and at the same time get it to both the top and bottom surfaces. We know that copper does the best job of conducting heat, followed by solder, and then FR4. Therefore, the best approach is to have a good thermal path between the power

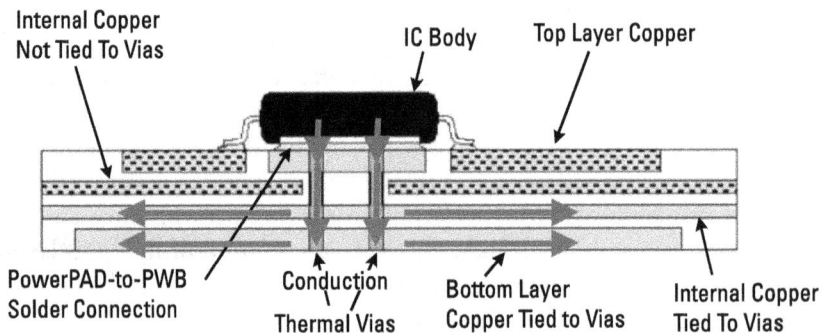

Figure 6.22   Thermal layout of a typical surface-mount power device.

pad of the device to at least one copper plane, either directly connected to the copper plane or by vias, and then let the copper plane spread the heat out. The larger the connection area to the heat source, the lower the temperature differential. Once the heat is spread out, it can be easily be conducted through the FR4 to the adjacent surface and then dissipated into the air.

Consider a two-layer PCB with a quad (2.5-mm × 0.3-mm power pad) surface-mount IC and a power pad. If thermal vias are not used, the heat has to be conducted through the FR4 material, to the copper on layer 2. For FR4 to be a fairly good thermal conductor, the layer has to be thin and there has to be a fair amount of area where the heat is applied. The theta value through the FR4 is $\theta = 794°C/W$, which is unacceptable. The only practical solution is to add vias in the power pad to lower the theta value through the FR4 material to the bottom copper layer. Now consider using six 13-mil thermal vias. The theta value drops to 16.3°C/W without solder filling or 13.4°C/W with solder filling. These six vias allow about 50 times more heat to be transferred away from the power pad of the IC to the bottom layer. The bottom copper plane spreads the heat out, dissipates heat off its surface, and also transfers heat back to the top layer where it can be dissipated.

In summary, thermal PCB design is a critical component design that is often overlooked or not understood. The ultimate goal is to keep the junction temperatures of the ICs and power devices below their maximum temperature values. The IC or power device typically has a theta value that accounts for the temperature rise from the power pad to the IC's junction. The board area and power dissipation determine the baseline temperature of the board, assuming there is at least one copper plane connected, by copper, to the IC. Multiple planes with multiple vias will help reduce the theta values. A larger package for a given heat dissipation provides a larger initial path and less temperature rise.

6.6 Assuring That an Intended Battery Is Used with the Device: Authentication

Let's assume that all considerations discussed in the preceding sections are followed and a safe battery/protector/cell-balancing/charging/system combination has been designed. All charging voltages would be set correctly, the safety threshold optimized, battery energy and power capability adapted to systems needs, and so forth. What can go wrong? Well, for one, this particular highly optimized battery might not be used at all with the system! Instead, for a mass market of millions of devices, there is a serious incentive for companies lurking in a gray area to introduce their own replacement batteries that physically fit into the space in your device and electrically fit your connectors. They would be able to sell these counterfeit batteries at a large discount compared to the original batteries because they are not following the same level of safety protections and might be skipping a particular functionality or maybe even all functionalities. For example, they might skip a primary or secondary protector, replace protection FETs with traces, use a constant resistor instead of a thermistor, skip the gauging IC, or use a sense resistor that has the wrong temperature coefficient. All of this can create various levels of discomfort for end users, from the occasional unexpected shutdown of the system or otherwise erratic behavior (that will be blamed on you as the original manufacturer) to a disastrous safety event that can cause significant damage both to the unsuspecting (or just uneducated about consequences) end user as well as to your brand name.

The simplest way to protect a system from counterfeit batteries is to make the connectors and battery itself with some intricate nooks and crannies that need to fit exactly into the counterparts of your system. However, in this age of rapid prototyping, protecting your system by just making an intricately shaped connector can be defeated with just ten minutes of work on copying the external shape of the battery and creating an identical form. Because "secondary market" manufacturers have to manufacture the casing for their battery using similar equipment to that of the original battery anyway, it does not make any difference to them if they have to make it match the exact shape or your battery or any other shape. The cost difference is negligible.

To raise a degree of difficulty in copying your original battery, an IC can be introduced in the battery that is supposed to report some expected response upon a system request. The system should be able to discern a valid response from an invalid one. The supply of such ICs would need to be tightly controlled by the original manufacturer, and the hardware design of such ICs should be difficult to copy. Several different approaches are currently available with increasing degrees of difficulty to circumvent them.

The simplest authentication method is to use an ID chip. The system sends an "ID request" command; the ID is reported by the chip and compared by the system with the list of valid IDs. Such a system is not too difficult to defeat, however, because the ID is communicated over the single-wire interface, which is easy to track with an oscilloscope. The communication protocols used are standard, such as HDQ as SDQ, so they can be easily decoded. Once the ID string has been decoded, a microcontroller that will produce the same response as the ID chip can be placed in the replacement battery, thus allowing it to pose as a valid battery. However, this scheme is still somewhat effective because it places the burden of adding an expensive microcontroller on a gray-area manufacturer, who is first of all trying to make the battery as cheaply as possible and second might not have the expertise necessary to complete the above operations. This added burden might make it no longer lucrative to manufacture counterfeit batteries (especially if battery itself is cheap; on the order of $5 to $15), while the cost of the ID chip to the original manufacturer is small, typically below 10 cents each in large quantities.

For more expensive batteries, a gray-area manufacturer might think it worth the trouble to create an ID-faking circuit. Therefore, stronger protection against an intercept attack is needed. A slightly more difficult to defeat option is a system that does not transmit the secret itself, but uses it to create a valid response. For example, the system can send a random data "challenge" to the authentication IC. The IC then uses the secret key to perform a transformation on this random data, creating a response that does not reveal the secret key itself, but is unique. In cryptography such a transformation is called a *hash transformation.* The system also stores the secret key and performs the same hash transformation on the random challenge data to create the "true response" data. Once a response is received from the authentication IC, it is compared with the "true response" and if they match, the battery pack is authenticated. Note that intercepting the response to one random challenge does not help when trying to compute the response to a different challenge (i.e., the secret key is irrecoverably "hashed" by the function), and the same challenge will never occur twice, so monitoring of a communication line does not help when trying to fake battery authentication. Figure 6.23 outlines the challenge–response authentication procedure.

Note that the hash function is more or less secure depending on which particular function is used. The simplest hash function is a cycling redundancy check (CRC) function. A CRC is often used to verify that a particular dataset

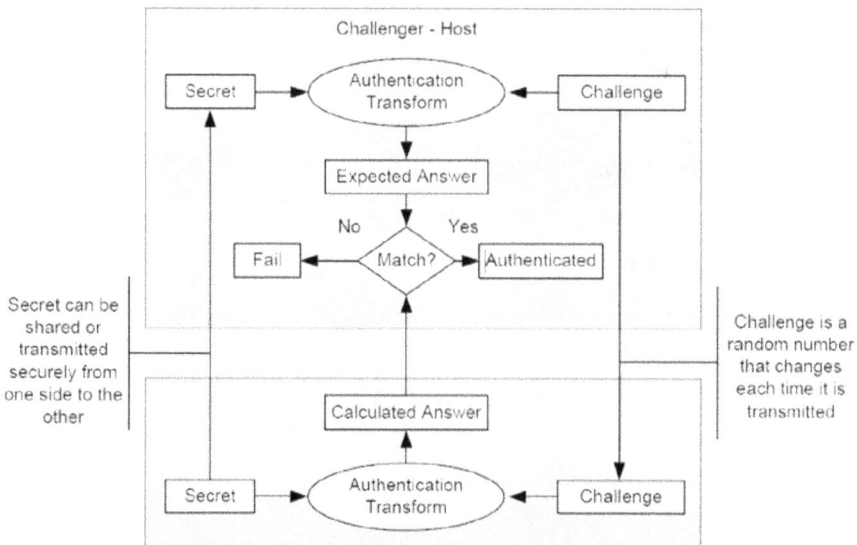

Figure 6.23   Challenge–response authentication technique.

is exactly the same as the one for which a check-sum was computed. It is used primarily for error checking of messages. Some implementations of authentication, such as the bq26150 IC, are based on the CRC function, and they compute a CRC check-sum from a secret key prehashed by a polynomial function with random coefficients and the challenge data to create a hash-response. Due to simplicity of CRC computations and small amount of memory required, such an IC would be the cheapest option for a somewhat protected authentication procedure. However, CRC-based hash functions have not been verified to be cryptographically strong, so their ability to resist cryptographic attack is questionable.

For more expensive battery packs, such as those used in laptops and tablet computers that constitute a higher temptation level to counterfeiters, it is advisable to go to a higher level of security that is based on standard cryptographic methods that have been proven to be unbreakable given reasonable computing time at the present level of computational abilities. Usually such methods remain valid even as computational capabilities increase as long as the secret key size is increased following the pace of computational advances, so that the overall architecture can remain unchanged. One example of such a standard secure authentication protocol is SHA-1/HMAC. Given a 128-bit secret key, this protocol is unbreakable given present computing capabilities. One example of an SHA-1-based authentication IC is the bq26100. A battery pack implementing this protection IC is shown in Figure 6.24.

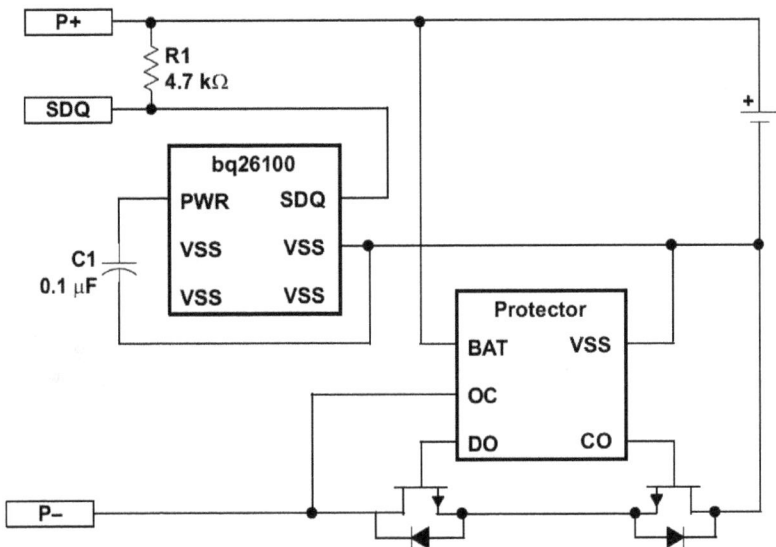

Figure 6.24 Example of a battery pack with a protector and authentication using a dedicated SHA-1 authentication IC (bq26100).

A cheaper way to implement SHA-1-based authentication for systems that already have a pack-side gauge IC is to use the IC's authentication capabilities. Examples of gauging ICs that support SHA-1 authentication are the bq27541 for a single cell and the bq30z55 for multiple cells in series. Such ICs are also available from Maxim, for example, the DS2784. Handling the authentication keys in this case is somewhat more complex because the battery pack maker generally has access to the data-flash settings of these devices until they are sealed. In this case it is important to establish a procedure during manufacture where the secret key is not programmed until immediately before the device is sealed. Much better security is provided by gauges with memory that is accessible only during the IC manufacturing process (secure memory); in such a case, the handling of the secure key remains in the hands of the IC manufacturer and it is not readable even by the pack maker. Examples of gauges with secure memory include the bq27545-G1 IC.

Another important consideration is the security of the silicon itself. While cryptographic authentication protects the communication intercept from compromising the security (e.g., software security), the chip itself is still exposed to possible hardware attacks in the form of attempts to duplicate the chip if the volume and cost of the counterfeit product are large enough to justify expenses such as reverse-engineering and the manufacture of an actual dedicated "fake" IC. Indeed, it is possible to analyze the chip under a microscope, see all preprogrammed logic gates and create a copy with the same switches ON and OFF, thus duplicating its operation. To prevent such duplication, authentication ICs have to be designed in such way that all of the information-carrying parts are covered with multiple layers of metal mesh. In addition, other techniques should be followed to make them hidden even from observation with an X-ray rather than just a simple optical microscope. While making an IC absolutely secure from a hardware attack is impossible, it is quite manageable to achieve a level of complexity that is out of reach for most companies except for the largest semiconductor players, which would not engage in such counterfeiting practices.

Another vulnerability to any authentication scheme is the handling of secure keys. With secret key-based hash functions, the key has to be stored both on the system side and inside the battery pack. A battery pack-side key would be controlled by the authentication IC maker (usually a large semiconductor company), which generally works within a controlled environment and has special key handling procedures in place that guaranty single-point responsibility for the key handling. It is more difficult to create such a system on the system side. If authentication is implemented in the software of the system's main processor, it is accessible to many different development teams, manufacturing teams, and so forth, so the number of people with access to the key is greatly increased. Secure memory programming during central IC manufacturing without any read possibility afterward is highly recommended in such designs. If secure memory is not available, another approach is to add another authentication IC on the device system side and use that to authenticate the battery pack. This method of key handling is simplified at a small added cost.

Another approach to make key handling more secure is to use public key cryptographic authentication, which requires only one secret key that would be stored in the authentication IC (which can be securely handled and stored); then "checking" if pack is authentic by the system is done with another key known as the "public" key. This public key does not have to be hidden in any way because it cannot be used to generate a valid response to the challenge, but just to check the correctness of the response. Public key cryptography is widely used in authentication of various transactions on the Internet. The most commonly used public key cryptography methods are RSA and elliptic curve cryptography. RSA has the benefit of being open source (patents are expired), and the computations needed for generating the hash data are not very extensive and can be performed in modern gauging ICs. One disadvantage is the large key size (at least 1,024-bit now and 2,048-bit to make it future-proof) required to achieve the security needed given modern computing ability. Elliptic curve cryptography (ECC), on the other hand, requires a much smaller key size (224 bits) for the same security level. However, many ECC implementations are still patent protected, and the massive computing power and computation time needed to compute the hash make it—for the moment—out of reach in terms of a reasonably priced authentication IC for battery applications, although such ICs in general are commercially available, for example, the ORIGA SLE95050 from Infineon.

References

[1]   "BQ20z70 and bq20z90 Applications," SULA404, Texas Instruments, 2006.

[2]   Jackson, B., "Battery Circuit Architecture," SLYP087, Portable Power Design Seminar, 2004.

[3]   Mammano, B., and B. Carsten, "Understanding and Optimizing Electromagnetic Compatibility in Switching Power Supplies," TI Power Supply Design Seminar, 2002.

7

Design Examples: Complete Battery Solutions for Specific Portable Systems

7.1 Introduction

In previous chapters we identified the different ingredients of a successful battery-operated system. Just to refresh in your memory, they include:

- Battery type and size selected to match the application.
- Protection circuit that ensures battery safety for one or more cells in series.
- Cell-balancing system.
- Remaining run-time indication system that allows a controlled shutdown.
- Charger circuit correctly sized for the battery capacity and required charge time.
- Robust ESD and magnetic interference protection.
- Thermal design that minimizes cells degradation.
- Authentication system that ensures the system is actually using the battery you designed for it.

In following sections we will apply these principles to specific portable systems to demonstrate them in action and to provide you with an actionable "reference template" that you can use as a starting point for your own development efforts for such systems or expand to other systems with similar properties.

7.2   Cell Phones and Smartphones

Today's smartphone integrates features such as personal organizers and e-mail management tools with basic multimedia and handset capabilities. For these reasons the ability to keep the phone running is more crucial to end users than it is for an entry-level mobile phone. The feature set drives the need for high processing power, larger and brighter displays, touch-screen controls, and speaker phone capability. To meet the performance and form factor requirements of the phone, careful consideration of the supply current and package size is important. Because the battery run time and cycle life are of serious concern to the end user, the battery management solution is crucial.

7.2.1   Battery Selection

Battery selection for a phone system is determined by the following main challenges.

High energy is needed to ensure a run time of 24 hours or more. This becomes especially challenging for smartphones because their loads are continuously increasing as a result of the new functions and apps that are constantly being added and that run at random times. The need for thin and lightweight phones became an important feature that is being compared among competing models. High current pulses applied by communication protocols such as 3G, CDMA, and LTA can cause the battery voltage to drop below the system shutdown voltage, which means that the power capability of the battery cannot be sacrificed to increase the phone's energy.

Basic functionality cell phones became a commodity; the are often available free with a cell phone contract and are extremely cost sensitive. For these types of units, manufacturers commonly try to find the cheapest possible Li-ion chemistry even if energy density is sacrificed. (Usually these types of cell designs are two to three generations older than the current generation.) Fortunately, the run time is still sufficient even when smaller batteries with less energy density are used because the functionality is limited to calling and the small display is less power hungry.

In contrast, for smartphones battery costs are less important than for a basic cell phone because the high price of the smartphone itself means that the battery makes up only 5% of the overall cost. Smartphones must have a very high energy density to support the run times required by these power-hungry multiple-feature phones, which points manufacturers toward choosing the battery with the highest energy density available, regardless of the cost.

The upper voltage range of smartphones is usually limited only by the 5V USB charging capability requirement, so any existing Li-ion chemistry would satisfy this range. This requirement, however, forces us to use a single serial cell configuration, since all high-energy chemistries would be above 5V for two or more cells in series.

Some older phones used NiMH batteries with multiple cells in series; however, that is becoming a rarity even for "free" phones because the energy density of NiMH is about three times lower than that of Li-ion chemistry. Nobody likes to carry around a heavy phone. The lower voltage range of cell phones is typically limited to above 3.4V due to the need for operational amplifiers for the RF transmitters. Because of their lower cost and higher energy, Ni-rich cathode batteries such as NMC (nickel, manganese, cobalt in a 1:11 ratio) and NCA (nickel ~80%, cobalt ~20%, aluminium ~0.1%) have recently become quite popular with cell manufacturers and notebook users. However, the above-mentioned high termination voltage makes their use impractical in cell phones due to their "depressed" discharge curve during most of the discharge process. This lower voltage, especially during the final portion of the discharge curve, would cause the system to terminate too early and to sacrifice too much capacity at high termination voltages as can be seen in Figure 7.1.

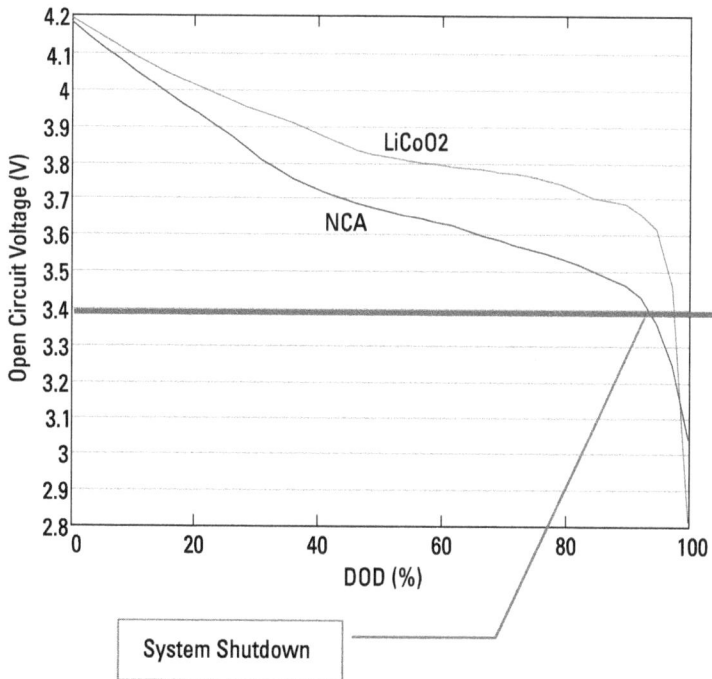

Figure 7.1   Comparison of LiCoO$_2$ and NCA voltage profiles in relation to a 3.4V shutdown voltage.

Using a buck-boost converter to maintain the system voltage at an acceptable level would allow us to extend the choice of available batteries with a potential cost reduction, but the efficiency of the boost has to be high enough to assure that a significant amount of energy is gained to justify its added cost and the space required for the converter's inductors and FETs. Currently, such buck-boost DC-DC converter systems are not considered to be cost effective, but the upcoming arrival of Li-ion cells with Si-based anodes, which will extend the "low-voltage energy" portion even further, will shift the balance toward using buck-boost systems.

Until that happens, we are left with the traditional $LiCoO_2$-based Li-ion battery (with highly crystalline anode and cathode materials to increase the volumetric energy density) as the only choice for smartphone batteries due to the very sharp drop-off from 3.65 to 3V, which makes the difference between capacity at a fully discharged state (3V) and actual system shutdown (3.4V) less than 2%. To further increase the energy of these batteries and to take advantage of newly developed higher voltage electrolytes and ceramic-coated separators, $LiCoO_2$-based cells are now often capable of charging to 4.35V, which has become common for smartphone batteries.

Regarding the type of battery package, using polymer pouch batteries is preferable if the battery is not removable, because this type of battery provides for the maximal possible energy density by weight and volume because it does not have a metallic casing. The phone casing itself should be designed to provide sufficient mechanical protection to the battery; maybe this is one of the reasons why nonremovable battery-operated devices often have a metallic casing. Even if the battery is removable, it makes sense to provide it with a short-circuit and overvoltage protector so that no short circuit or other mishandling occurs during assembly of the device.

For a removable battery, a prismatic cell is the only choice since the thickness of the device is limited by user convenience. Standard cylindrical cells, like the 18650 cells, are out of the question, and the polymer pouch cell is not robust enough to be handled by end users.

Battery size selection is determined by the energy needed to achieve the required run time and the power capability needed to ensure that the highest current spike expected in the system will not cause the battery voltage to drop below the system shutdown point. The latter factor is very important for phone systems because very high current spikes can occur during transmission. Typically the cell is sized to ensure that no more than 200 mV of voltage drop is observed during a transmission spike. Note that a short-pulse response voltage drop needs to be added to the termination voltage. For example, if a system will shut down at 3.2V, but the spike is 200 mV, then the effective termination voltage for steady-state discharge (without a spike) becomes 3.4V. A higher spike would move the effective termination voltage to a value above 3.4V, which can make system operation unstable since the flat portion of the voltage curve then comes too close to the termination voltage, and the usable capacity dependence on the rate of discharge and temperature increases dramatically, making the run time inconsistent.

7.2.2 Battery Pack Electronics

To ensure the basic safety of cell phone battery packs, and industry-wide standard, IEEE 1725, has been established. It defines the generic requirements for battery pack safety as well as quality control and reliability. There is no general standard for communication with a battery pack due to the high diversity of cell phones, although a few companies in the field with the largest market share have established their own standards with which most users are familiar: communication protocol HDQ for single-wire and I²C for two-wire communication. Two-wire communication is most likely used during production to speed up the flash programming process, whereas single-wire communication is used in the final battery pack to reduce the pin count. Maxim uses SDQ for single-wire communication. Communication between the gauge and a PC that is used to configure it and monitor various tests that can be run on a battery pack is accomplished using a USB-to-I²C or USB-to-single-wire interface. While communication protocols are open source and users can develop their own communication devices, IC manufacturers usually provide such interface devices as well as control software for the PC with their evaluation boards. For example, TI provides the EV2300 USB/I2C/HDQ interface and evaluation software, which allows for the monitoring or logging of all gauge-reported values, the changing of data-flash settings, and the performance of calibration or test authentication processes. The command set that can be used is not standardized across the industry but usually is common between each IC manufacturer's devices. In the case of TI, the common command set for all single-cell battery gauges includes battery state values that can be read from a gauge by a host. Values include remaining capacity, full capacity, state of charge, remaining energy, state of health, voltage, current, temperature, and values used for analysis and testing such as charging status and control status indicating the different stages of battery operation.

A single-cell battery pack configuration depends on the intended functionality of the phone. For basic phones, just a protector is sufficient. SOC information is provided by a host-side "function counter" that subtracts a certain amount of energy per function or by simple voltage correlation gauging. Obviously such an approach is in error on the order of ±25%, as discussed in detail in the chapter on gauging, but it is sufficient as a simple warning to the user that it is time to plug in the charger. Since no information processing is done in a basic phone, the soft shutdown capability with data preservation is not required. For smartphones, a battery gauge is either added to the battery pack or on the system board. Figure 7.2 shows a schematic of a smartphone battery pack including the battery gauge, protector, and authentication IC residing inside the battery pack. There are also ICs that provide both gauging and protection, such as the bq28550.

For accurate coulomb counting, a 10- to 20-mΩ sense resistor is used, preferably with a low temperature coefficient for best accuracy over a temperature range. The thermistor should be placed close to the surface of the cell. Since a smartphone battery pack is very compact, placing a thermistor on the PCB together with other components is close enough for most cases. The protection IC should be chosen so that the overvoltage threshold is about 50 to 100 mV above the charging voltage to account for charger tolerance while still providing sufficient protection. The undervoltage threshold can be set slightly below the recommended discharge termination voltage to account for IR drops during current spikes that drive the effective voltage much lower than OCV. That would cause discharge termination to occur too early if the cell manufacturer recommended minimal discharge voltage is used. Using a 2.7V undervoltage limit for a cell with a recommended 3V termination voltage is reasonable

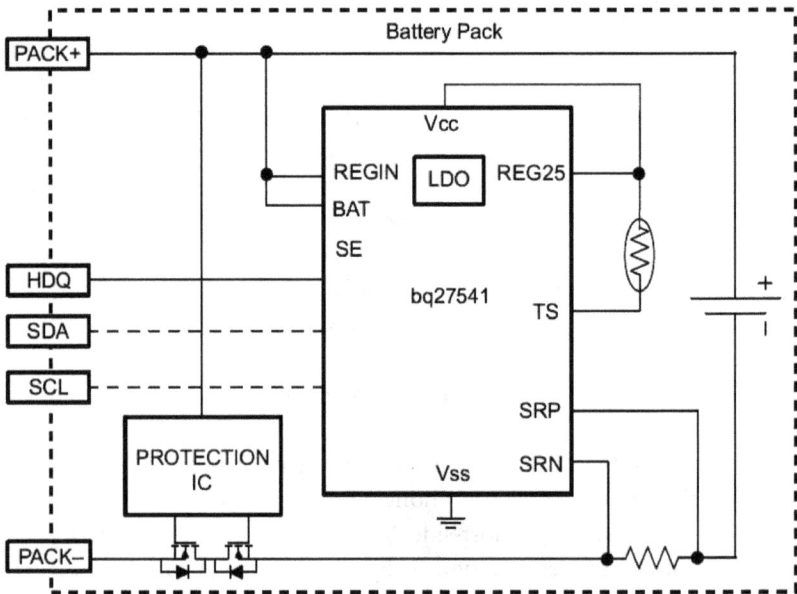

Figure 7.2 A smartphone battery pack with a battery gauging/authentication IC (bq27541) and a protection IC.

for a cell phone due to its spiky load during transmission. A few millisecond duration of undervoltage during a spike does not damage the cell because an electrochemical reaction will not start before all of the short-time-constant processes such as recharging the double layers are finished.

Electromagnetic interference (EMI) with a gauging circuit operation is very possible in a cell phone due to presence of a powerful transmitter. For this reason, it is critical for the high ohmic traces from the sense resistor to the gauging IC to be as short as possible and to provide a filter network as shown in Figure 7.3. Voltage measurement traces also require a low-pass filter with a low-pass frequency set to half of the sampling rate of the data convertor used for voltage measurement.

Thermal design of a cell phone includes keeping hot elements from coming in direct contact with the casing because that could cause discomfort to the end user. Unfortunately, doing so means that the battery is acting as a "heat spreader." This is not ideal in terms of battery degradation (keeping it far from any heat sources would be best for the battery), but if we have to choose between protecting the battery or the people we have to reluctantly choose the latter.

Since cell phones and smartphones are produced in huge quantities, the incentive for third parties to develop a compatible battery pack and try to sell it instead of the original battery pack at a discount on the black market is quite high. These replacement batteries can be dangerous if they have the wrong charging voltage, do not have a protector, or use cells from a not very safety-conscious manufacturer. They would be prone to causing internal short circuits. To protect the end user from using such a questionable device, as well as protect yourself from tremendous damage to your brand name in the case of a safety event, it is useful to set up an authentication system. An authentication system will allow the phone's system to verify that the battery is original and, if it is not, to take action so that the replacement battery cannot be used. Actions can range from giving the end user a warning message to disabling the charging process. Any of the authenticating methods described in Section 6.6 are suitable, with more sophisticated methods making more sense for higher end models. The easiest implementation is to provide authentication in the gauging IC, because it comes as a free bonus. For example, the bq27541 shown in the schematic in Figure 7.2 provides SHA-1 random challenge/response authentication that is much more secure than most credit card transactions in the United States (which are just plain ID based). If more security is required or host-side gauging is used so there is no gauge IC in the pack, a separate authentication IC such as the bq26100 (also SHA-1 based but with added hardware-based security) is used. Equivalent ICs are available also from other manufacturers such as the DS2703 from Maxim and the ISL9206 from Intersil.

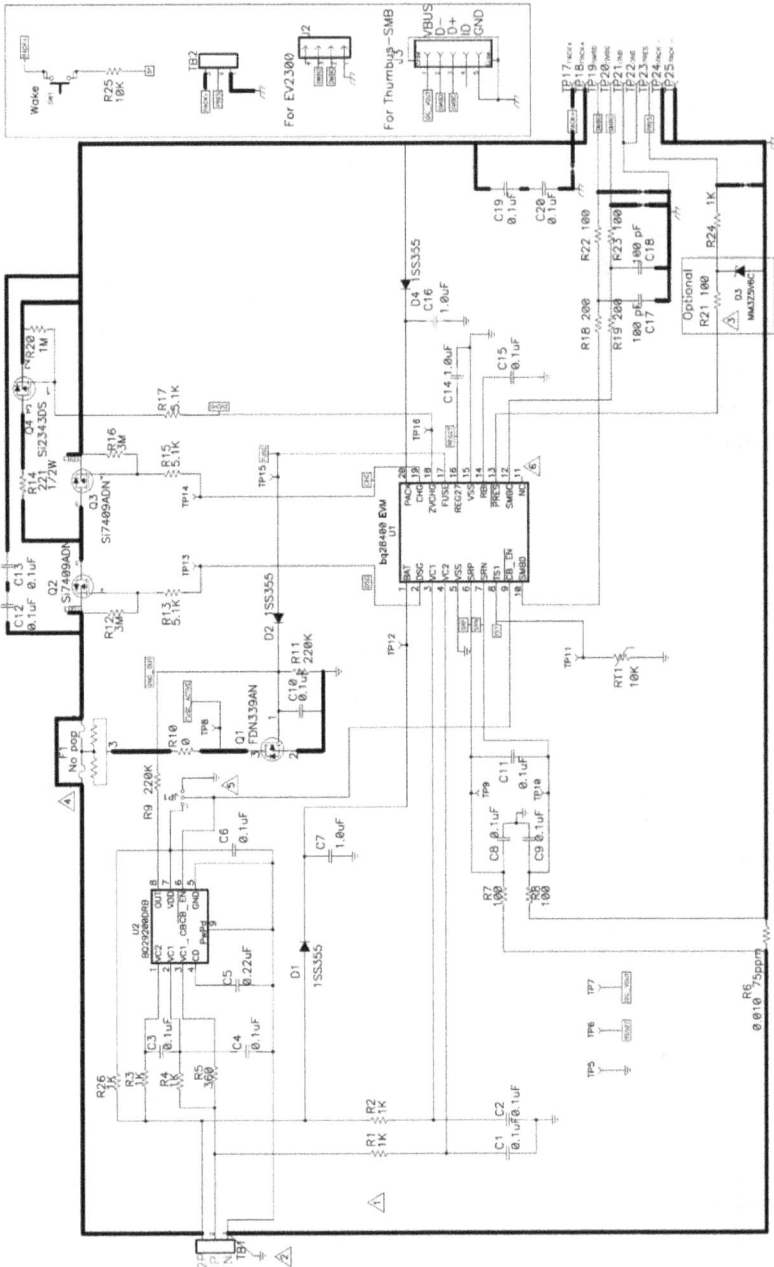

Figure 7.3 A schematic of a battery gauging and protection solution optimized for a two-cell serial configuration for a tablet PC, based on a bq28400 gauge/protector IC and bq29200 secondary protector IC.

7.2.3 Battery Charging

Smartphone end users typically like to have fast charging and want to have 80% SOC in 1 hr, which means that the battery can be charged in the morning or during lunchtime and then used for the whole day. This requires the battery to have a fast-charging algorithm through either a high charge current or other unique charging algorithm. Most smartphones currently have 1,500- to 2,000-mAh batteries, and usually a 2A charge current is high enough. However, with the improvements in battery chemistry and the larger phone displays (e.g., 5 in.), some smartphones have started to use charge current greater than 3A to achieve fast charging.

Most of the phones use the same physical connectors such as the MicroUSB for both USB ports and the adapter. The charger typically needs to detect whether the power sources are from a USB port or adapter, so that we can set the maximum input current limit to avoid a power source crash. In addition, the system power required by the applications processor or power amplifier is very dynamic with a very high pulsating power. The power from an adapter or USB port may not be high enough, which would require the battery to provide temporary power to support the system. Therefore, the NVDC battery charger is usually needed. For a given limited space, the smallest size solution is definitely preferred by operating the charger at a multimegahertz switching frequency so that a chip inductor can be used. In most smartphones, the PMIC usually integrates a switch-mode battery charger up to 2.0A due to thermal concerns. For higher than 2.0A single cell battery charger, an independent battery charger is usually considered for achieving fast charging. Here are some of the most popular chargers like SMB347, bq24161 with 2.5A charge current, and bq24190 with up to 4.5A charge current in smart-phone applications.

7.3 Tablet Computers

Tablet computer systems are similar to smartphone systems in the sense of comparable computing power and RF communication support. The main difference is that tablets have a much larger display size and resolution and a corresponding increase in power consumption by the display backlight and the more powerful graphic processing engine. Also increased is the run-time requirement for continuous usage, because tablets are often used as a continuous book reader, video viewer, gaming device, or even work tool. Increasing the run-time requirement and the energy required by the display has an interesting effect—since a tablet battery is already very large, but communication requires the same current as is used in smartphones, the power capability of a tablet battery is much higher than needed to handle the communication pulses, and voltage drops during communication spikes become much smaller. This allows us to decrease the effective termination voltage compared to smartphones (even if other parts of the system electronics remain same), which will make the run time more independent of load and temperature variations. In addition, the cells' internal design can be optimized to be more of an "energy" battery than a "power" battery, which could add some more energy density that is sorely needed.

Since the backlit display is responsible for a large portion of overall power consumption, it makes sense to optimize the battery system for higher efficiency of the power rails used for the backlight. Traditionally the cold cathode fluorescent light (CCFL) backlight was the highest voltage part of the overall electronic system—it requires voltages from 800 to 1,300V. Since this is much higher than any battery can provide, voltage boosting systems have to be used anyway, but it helps the efficiency if the base voltage is higher. In traditional laptops the battery voltage is kept at 16 to 12V, which requires three to four cells in series. But tablets are sized such that they can be held in one hand or at least comfortably in two hands, so they need to be much thinner than laptops. From this point of view, using cylindrical cells (which are the most cost effective) is out of the question because they would not fit. It might be possible to use multiple prismatics, but the use of too many cells side by side in a flat configuration is not very mechanically robust and would require more supporting structures. Using multiple cells also occupies valuable space because multiple casings are required. For this reason, despite the benefits of having higher voltage, most tablet systems use a single cell, like smartphones, or maximally two cells in series.

From the display power rail standpoint, future designs are likely to move to lower pack voltages because backlights are moving from CCFL to the less voltage needy LED backlight (which usually require 2.8 to 18.5V) or even the OLED (2 to 4V). Using one cell in series has the added benefits of being able to use the same 5V USB charger connector as smartphones are using and being able to charge from the USB port of a computer. However, a USB port is only guaranteed to provide 500 mA, which is very small for the huge battery used in tablets and would require charge times of 15 to 20 hours. For this reason USB charging as such is less beneficial for tablets, so having a higher voltage to provide better efficiency of the backlight power rails may shift the decision to using two cells in series and forgoing the USB charging ability. Pressure to go to two cells in series will also come from the battery chemistry development road map—cheaper Ni and Mn containing cells with a lower voltage profile at low SOC (but higher overall energy), as well as the upcoming even higher energy density Li-ion cells with Si-contained anodes that also add energy mostly below 3.5V, will favor stacking the cells in series to make termination voltage per cell lower than 3V while still being able to operate the system efficiently.

7.3.1 Battery Pack Electronics

Because a one-cell battery pack configuration is identical to that of a smartphone, refer to Section 7.2.2 for this information. In particular, a schematic of a battery pack is given in Figure 7.2. For two cells in series configuration, the battery pack design becomes more similar to that of a notebook (see Section 7.4.2), so any two- to four-cell serial configurations with gauge/protector solutions such as the bq30z55 IC become an option. For lower end systems with severe limits on battery cost, sometimes ICs dedicated specifically to a two-cell serial configuration are used such as the bq28400. These ICs also offer a somewhat smaller reduced solution size and part count compared with traditional notebook gauges, while maintaining the same smart battery specification (SBS) command set. A typical implementation schematic is shown in Figure 7.3.

As can be seen from the schematic, the bq28400 IC provides the gauging function, communicated to the host using a two-wire SMbus (basically I_2C with a more narrow specifications). Two cells in series are connected to 1N (cell −) and 1P terminals and 1P and 2P terminals, correspondingly. Note that in addition to first-level protection provided by the bq28400 that controls charge and discharge FETs Q2 and Q3, an additional independent protector, the bq29200 IC, is provided. An independent overvoltage protector is required for all configurations with more than one cell to avoid a catastrophic failure in the case of primary protector malfunctions.

In addition to a protection FET, the additional mechanism of cell disconnect is also used in all multiple-cell packs. This mechanism is referred to as a chemical fuse or three-terminal fuse (F1 in the schematic of Figure 7.3). It is an unusual type of fuse that disconnects the circuit not under an overcurrent condition, but rather when external voltage is applied to its third terminal. Only a small current is need to blow the fuse, so it can be operated even under short-circuit conditions when the battery voltage drops to a low level. The purpose of this independent protection is to be able to disconnect the cells even if protection FETs themselves have failed short, those providing an additional level of redundancy to the shutdown mechanism. The gauge can detect FET failure as well as some other self-check failures or internal cell failures, such as excessive imbalance, and permanently disable the pack by blowing the fuse and preventing unsafe situations that can be caused by lost protection levels. Note that both the primary and secondary overvoltage protectors can blow the three-terminal fuse independently to ensure that cell overvoltage never happens even if everything else fails.

As usual, temperature information from a thermistor, as well as voltage readings across each cell and current readings indicated by a voltage across a sense resistor, provides information used for safety and gauging decisions. In Figure 7.3, note the filter network protecting the kelvin connection across sense resistor R6 from possible system noise. A successful PCB layout should place corresponding capacitors and resistors as close as possible to the IC pins.

7.3.2 Battery Charging

Please refer to the detailed discussion of battery charger design examples for tablet computers in Section 2.5.1.

7.4 Notebook PCs

Notebook PCs have been leading the development of sophisticated power sources due to their extreme power consumption and small size requirements. Notebook PCs were expected to deliver effectively the same performance to the user as a desktop PC that is usually 5 to 10 times larger in size and weight and is also connected to an AC grid, and also last 2 to 3 hours on their own portable power supply. Unreasonable as this expectation has been, it was successfully met by the industry for years, resulting in a notebook PC being a true marvel of electronics and mechanical engineering. The main reason for this success is the high energy density provided by Li-ion batteries. However, this energetic beast needed to be tamed so it could be safely stored not only in NASA secure facilities but also in people's homes, where it may literally lay on beds beside sleeping babies, and in packed airplanes crisscrossing the skies. To achieve this extremely safe behavior while dealing with the added complexity of multiple cells in series, designers needed to achieve the high voltages common in PC architectures, making the notebook battery pack the most complex of all battery packs used in portable electronics.

7.4.1 Battery Selection

Because a notebook PC usually operates with two to four cells in series, cell minimal discharge voltage is usually not a limiting factor because systems can operate in most cases down to lower voltages than the cells can be discharged to. For example, for three cells in series, 3V/cell corresponds to 9V/pack, but systems can usually operate down to 7V. This opens up the possibility of using chemistries different from $LiCoO_2$ since most of the "low-voltage" energy of the Ni- and Mn-based chemistries can be utilized. For this reason, NMC and NCA cells are widely used in notebooks along with traditional $LiCoO_2$ cells. This results in lower cost without sacrificing energy density because of the much lower price of Ni and Mn compared to cobalt. Upcoming Si-containing anodes can also be accommodated in notebook systems without any changes to the electronics.

Most notebook systems use cylindrical 18650 cells, where "18650" stands for the cell's dimensions: 18 mm diameter and 65 mm height. There is a significant range of energy densities and prices for 18650 cells. For example, in 2012, cells were available in the low (2,200-mAh), medium (2,600-mAh), and high-end (3,000-mAh) ranges. Of course, capacity is a moving target that has grown about 7% each year since the inception of the Li-ion battery. Prices vary roughly from $2 for a low-end cell to 3$ for a high-end cell, although these values are highly dependent on market conditions and the price of raw materials (e.g., cobalt), which tends to be highly variable. Prismatic cells that use a metallic casing are used more rarely for "thinner" designs where cylindrical cells will not fit. Prismatic cell sizes are much more variable, so sizes fitting any design can be found. However, because they are less standardized, they are likely to be more expensive than 18650 cells for the same energy. To obtain the thinnest designs possible, polymer pouch cells can be produced in any size and thickness and have the highest energy density by weight because there is no heavy metallic casing to carry around. If the battery pack is removable, however, the external casing for polymer cell-based battery packs needs to have much more mechanical strength and puncture resistance because the polymer pouch itself provides virtually no protection from mechanical abuse. Nonremovable designs are preferable if thinness and lightness are the main goals.

The particular choice of the battery will depend on the type of notebook PC system. Notebook PC designs cover a wide range of customer preferences and include the following:

1. High-end business PCs where fast charge times and long run times are critical. Performance requirements are high and expected to be on par with those of desktop PCs. They include the same high-end graphic engines and processors as PC systems (although slightly down scaled in frequency and possibly with a higher end, smaller element size silicon process to reduce power consumption). Such systems would need the highest energy density battery out there on the market, which is typically realized with $LiCoO_2$-based cells of the highest quality and use of advanced electrolytes that allow charging to 4.35V/cell. To ensure the longest run time, four series, two parallel (4s2p), eight-cell configurations are often used, although 3s3p and even 4s3p are sometimes employed for particularly energy-hungry systems. Business systems tend to use the high-end cylindrical cells and sometimes prismatic or polymer cells for the ultrathin models.

2. Medium-range consumer-oriented systems have more relaxed requirements for performance (lower frequency processor, less memory, often no dedicated 3D graphic card), but cost is critical. System are more likely to use Ni-containing cathode cells like NCA and NMC, and

capacities range from low to medium to achieve reasonable cost. These systems commonly use 3s2p, six-cell battery packs. Some high-end ultrabooks use polymer cells.

3. Low-end notebooks and their smaller size variety, called netbooks, usually offer performance that is good enough for social networking and web use, so no high-performance graphics, large hard drive, or large memory is included. A lower capacity battery can therefore be used. The 2s2p configuration is common in lower end notebooks. Some convertible tablets/PCs with detachable keyboards sometimes use polymer cells due to thin form factor requirements.

7.4.2 Battery Pack Electronics

Because notebook battery packs always include multiple cells in series, the pack electronics have to accommodate both individual cell voltage monitoring and cell balancing in addition to the overcurrent protection, capacity indication, and authentication requirements that we already encountered earlier in single-cell pack designs. Since notebook battery packs have high energy, two independent overvoltage protection circuits are needed to make an overvoltage situation virtually impossible. Another new element in notebook battery pack electronics is the possibility of controlling the charger residing in the system by means of a configuration settings (such as charging voltage and charging current) readout from the battery pack, which enables different types of battery packs to be used with one system.

Given the high complexity of notebook battery packs, an industry-wide standard, IEEE 1625, governs their design, and the smart battery specifications (SBS) define a communication protocol and a set of commands that notebook battery packs are supposed to support. The SBS command set is shown in Table 7.1. (Refer to the datasheet for each specific device for more details.) Use of the SBS command set provides the convenience of learning only one set of commands and arranging only one set of communication equipment even if different types of battery packs are dealt with, as opposed to the situation in the single-cell battery packs world where no single parameters standard exists.

The pack is usually "sealed" after manufacturing using a seal command, so that no access to data-flash and no modification of settings are possible without having the unseal code. Note that additional commands may be available depending on the particular gauging IC manufacturer when the pack is in the unsealed mode. (Refer to the datasheet for a particular device for possible extended commands info.)

Due to the complexity of notebook battery packs, they are exclusively based on specially designed gauging ICs that allow us to minimize the development time, yet ensure the highest safety and conformance with industry

Table 7.1

SBS Command Set

SBS Command Name	Address (Hex)	Meaning
ManufacturerAccess()	0x00	This command is followed by an additional 2-byte word and defines access to various commands used during manufacturing.
RemainingCapacityAlarm()	0x01	Sets alarm bit if remaining capacity is below a predefined threshold.
RemainingTimeAlarm()	0x02	Sets alarm bit if remaining run time is below a predefined threshold.
BatteryMode()	0x03	Reports bit collection with different battery mode settings, for example, a setting for reporting charge in mAh or cWh.
AtRate()	0x04	Sets the rate of discharge for which AtRate values will be reported.
AtRateTimeToFull()	0x05	Time to full that corresponds to the rate previously set by the AtRate command.
AtRateTimeToEmpty()	0x06	Time to empty that corresponds to the rate previously set by the AtRate command.
AtRateOK()	0x07	Indicates that values have been updated after the AtRate command changed the requested rate.
Temperature()	0x08	Temperature, K*10
Voltage()	0x09	Pack voltage, mV
Current()	0x0a	Pack current, mA
AverageCurrent()	0x0b	Average pack current, mA
MaxError()	0x0c	Estimated error of SOC reporting, %
RelativeStateOfCharge()	0x0d	SOC at present rate of discharge, %
AbsoluteStateOfCharge()	0x0e	SOC at zero rate of discharge, %
RemainingCapacity()	0x0f	Remaining charge, mAh or cWh depending on battery mode
FullChargeCapacity()	0x10	Capacity available at present rate of discharge, mAh or cWh depending on battery mode
RunTimeToEmpty()	0x11	Time to empty at present rate of discharge, sec
AverageTimeToEmpty()	0x12	Time to empty at average rate of discharge, sec
AverageTimeToFull()	0x13	Time to fully charge at average rate of charge
ChargingCurrent()	0x14	Current to be applied by the charger during constant current (CC) mode of operation
ChargingVoltage()	0x15	Voltage to be applied by the charger during constant voltage (CV) mode of operation
BatteryStatus()	0x16	Bit collection reporting various battery status bits such as fully charged, fully discharged, and so forth
CycleCount()	0x17	Number of cycles to which this battery has been exposed
DesignCapacity()	0x18	Predefined design capacity, mAh
DesignVoltage()	0x19	Predefined design voltage, mV
SpecificationInfo()	0x1a	{AU: Need entry here?}
ManufactureDate()	0x1b	Date pack was manufactured
SerialNumber()	0x1c	Pack serial number assigned during manufacturing
ManufacturerName()	0x20	Name of pack manufacturer
DeviceName()	0x21	Name of the device
DeviceChemistry()	0x22	Cell chemistry, if several are supported
ManufacturerData()	0x23	Arbitrary data string for information purposes

standards. Such ICs vary in the level of integration and supported functions such as LEDs. We can take a closer look at such a solution using an example for a full-featured notebook battery pack based on the bq20z65 gauge/protector, shown in Figure 7.4.

This design is similar to that discussed for the 2s tablet computer battery pack, with one difference being a larger number of cell connections. The same filtering on the cell connections and sense resistor kelvin connection is used. The I^2C communication lines are protected against ESD hits with Zener diodes D5. A chemical fuse-blowing circuit is also provided and brought out to the test point fuse (TP3); however, the fuse itself is not populated in this schematic because it is an evaluation module. Because notebook battery packs are usually removable, more functionality is sometimes added for user convenience, such as an LED that can light up to indicate the present state of charge upon the press of a button. SBS communication lines are protected.

7.4.3 Battery Charging

Refer to the detailed discussion of a notebook battery charger in Section 2.6.2.

7.5 Ultrabooks

An ultrabook is a high-end type of subnotebook defined by Intel as a new category of ultra-responsive, ultra-sleek, and ultra-stylish devices less than 20 mm thick. Ultrabooks are designed to feature compact size with extended battery life and without compromising performance. They use low-power Intel Core processors, solid-state drives, and a unibody chassis to help meet these criteria, and typically use a 45W or lower adapter. Due to their limited size, they typically do not have disk drives or Ethernet ports. Rapid start is another important feature; ultrabooks can access applications and frequently used files instantly by using SSD drives (flash memory) instead of a traditional mechanical HDD. Intel Turbo Boost Technology 2.0® allows processor cores to run faster by managing current, power, and temperature and automatically giving you a burst of speed whenever you need one, for instance, when opening Outlook to access e-mails.

7.5.1 Battery Selection

Due to the thin body requirement, polymer cells, which are often nonremovable, are a must. In addition, due to turbo-mode pulses, cells need to have low high-frequency impedance to be able to provide 10-ms-long 4C rate discharge pulses without dropping the voltage below the system shutdown level. Most common polymer cells can easily satisfy these requirements.

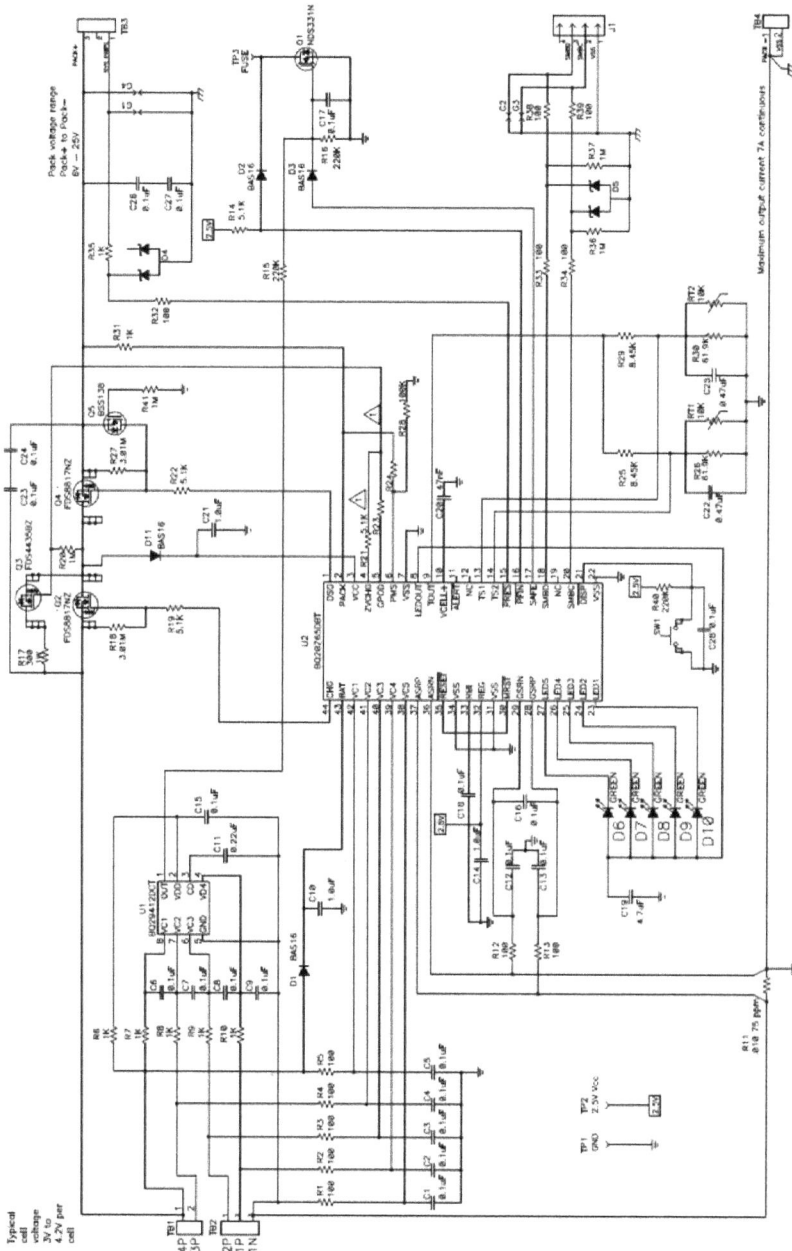

Figure 7.4 Notebook battery pack evaluation module using a bq20z65 gauge primary-level protector and bq29412 secondary-level protector.

7.5.2   Battery Pack Electronics

Battery pack design includes 3s and 2s configurations that look similar to those described earlier. Battery gauging requirements are somewhat different because for effective user of the turbo mode, the gauge should be able to report to the host the maximal power the battery can provide during a spike without dropping the voltage below the termination voltage point. If the system's presently required power exceeds the reported MaxPower, the system is able to throttle down its turbo settings to reduce the level of spikes. In this way the user can benefit from faster system operation during most of the discharge process, and still able to discharge the battery to empty without terminating too early on a high current spike. Of course, to achieve this function, the gauge should be able to model the battery's transient responses, in particular, its high-frequency impedance, which corresponds to a particular spike duration. For example, if a spike is applied for 10 ms, the effective impedance that corresponds to this duration (roughly 0.1 kHz) needs to be used for the MaxPower calculation. Fortunately, high-frequency impedance is mostly due to current collectors and separators and as such it changes very little with cell aging, as can be seen in Figure 5.11, so cell characterization data can be used for this computation. In addition to cell resistance, system and pack resistances also need to be considered.

The gauging system also needs to be aware of this power throttling by the system so that remaining capacity is reported under the correct consideration that the turbo-mode spike level is going to be reduced as end of discharge approaches. A good example of a gauge that is especially designed to support Intel's turbo mode is the bq30z554. Figure 7.5 shows an example of voltage and current during turbo-mode operation that is adjusted to ensure that the power stays below the gauge-reported MaxPower value so that voltage does not go below the termination point even during a spike.

We can see from Figure 7.5 that operation at maximal turbo mode (4C rate pulses) continues until about 0.4 hr is left, at which point the MaxPower capability of the battery becomes less than 4C as reported by the gauge, and after that it is gradually reduced by the system to achieve maximal system operation time. Note that voltage always remains above the termination voltage during operation because the gauge predictively warns the system about the available power limits and causes it to throttle down before voltage dropping below shutdown voltage during a spike could occur.

7.5.3   Charging and Power Architecture

The power architecture of the ultrabooks is similar to that of traditional notebooks. The goals are to achieve the highest power conversion efficiency, excellent thermal management, and the lowest profile solutions, which require all components to be less than 3 mm in height.

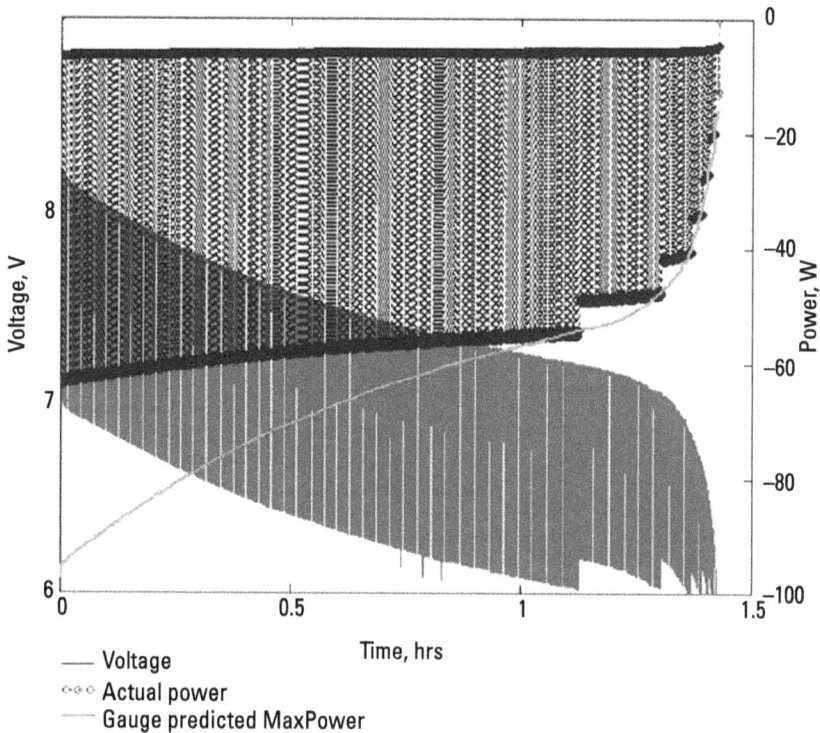

— Voltage
∘∘∘ Actual power
---- Gauge predicted MaxPower

Figure 7.5   Discharge at the C/2 rate with turbo-mode 10-msec spikes initially at the 4C rate occurring every 10 sec. The turbo-mode power level is reduced when the MaxPower value reported by the bq30z554 gauge is exceeded. Power is shown as negative, since it is a discharge.

Figure 7.6 shows the ultrabook's power architecture. The adapter output is connected to the battery charger input as a battery charging power source. The battery charger output powers the system and charges the battery. From this architecture, the system bus voltage is always close to the battery voltage, which is typically 6–8.4V for a 2s battery pack and 9–12.6V for a 3s battery pack. Therefore, it has a very narrow system bus voltage, which allows for the use of 20V voltage rating power MOSFETs with lower on-resistance to achieve very high power conversion efficiency with high-frequency operations in the megahertz range and minimization of the solution size. Therefore, it is not desirable to use a NVDC charger for a 4s battery pack configuration because its maximum battery voltage could be as high as 17.4 for 4.35V/cell.

Compared with the traditional notebook power architecture, the system bus voltage is either adapter voltage, typically 19.5V when the adapter is connected, or battery voltage of 6 to 8.4V for a 2s battery pack. Therefore, the system bus voltage varies from 6 to 19.5V, which requires use of 30V voltage rating power MOSFETs. It is difficult to operate at the 1-MHz switching frequency

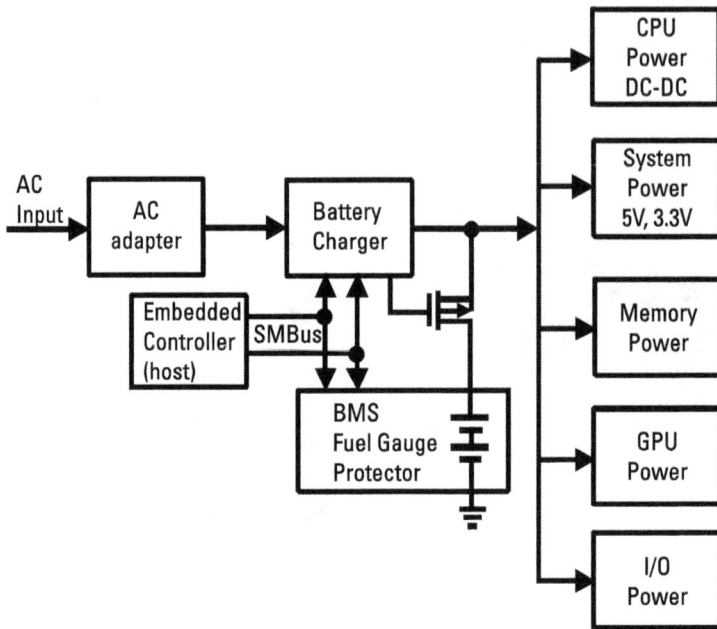

Figure 7.6   Ultrabook power architecture block diagram.

with a wide range of input voltages from 6 to 19.5V for minimizing the inductor size and optimizing the efficiency for the downstream DC-DC converters. Table 7.2 shows the benefits and challenges of NVDC chargers.

In terms of system considerations, there are a few challenges for a NVDC charger used in ultrabook applications:

- The height of all components has to be less than 20 mm, which requires high switching frequency operation (e.g., 800 kHz).

- In the Intel CPU turbo mode, the CPU peak power could be much more than the adapter power rating of about 10 ms. So as not to increase the adapter power rating, the system must support battery discharging for supporting the CPU turbo boost while the adapter is connected. This also implies that the charger has to have a very fast input current transient response to reach its input current regulation point without overshooting; this will help avoid an adapter crash during system and CPU step load transients.

- For a deeply discharged battery or absent battery, the system bus voltage should have a very fast transient response and its output voltage cannot be lower than 5.5V for a 2s battery pack so that the system can always operate.

Table 7.2

Benefits and Challenges of NVDC Chargers

Benefits	
Minimum number of output cap in Vcore	Lower system bus voltage, higher fs of 1MHz; smaller Vcore solution size
Improved efficiency in Vcore	Lower Rdson and Qg with 20V FETs than 30V FETs
Possible battery mode efficiency improvement	Lower Rdson and Qg of 20V FET, but increases loss due to charge current Rsense
Challenges	
Charger size and cost increases	Higher inductor and FETs current rating. But not significant for ≤ 45-W system
Energy star, or ERP6.0	Light load efficiency; PFM; lower Iq
Input current DPM transient	Overloading adapter. Fast DPM transient response

The NVDC charger is used not only as a battery charger, but also as a first-stage DC-DC power converter for converting the adapter voltage into the system bus voltage. The whole system is a two-stage power conversion system. When the battery is fully charged and system is in low-power mode, the battery charger is not off, but operates as a DC-DC converter. To meet Environmental Results Program (ERP) lot 6.0 in Europe, the total input AC adapter input power must be less than 0.5W in system standby mode. This requires the battery charger to have a very high light load efficiency at from 100 to 200 mW of output, preferably an efficiency of more than 80%.

7.5.4   Ultrabook Battery Charger Design Example

Battery pack: two cells in a series battery pack, 6,000 mAh.

Adapter: 19.5V/2.3A (45W).

Efficiency: minimum 80% at 6V/30 mA output.

System voltage transient response: ≤ 0.4V drop at 0 to 3A step load with 1 A/μs slew rate.

Figure 7.7 shows the NVDC charger design example. It has a selectable switching frequency of 600 kHz, 800 kHz, or 1 MHz through an SMBus control register for achieving the smallest inductor size. The Q1 is a can-MOSFET, which is used to control the inrush current when the adapter is inserted. Because decoupling capacitor C3 is typically small around 10 to 20 μF, Q1 may not be needed if the adapter can handle the inrush current. Q2 is a MOSFET for blocking the battery leakage current to the adapter side. Both Q1 and Q2 are N-MOSFETs designed to have lower on-resistance or be inexpensive.

To maintain the minimum system bus voltage of 6V for a deeply discharged battery pack, the MOSFET Q3 operates in a linear LDO mode for

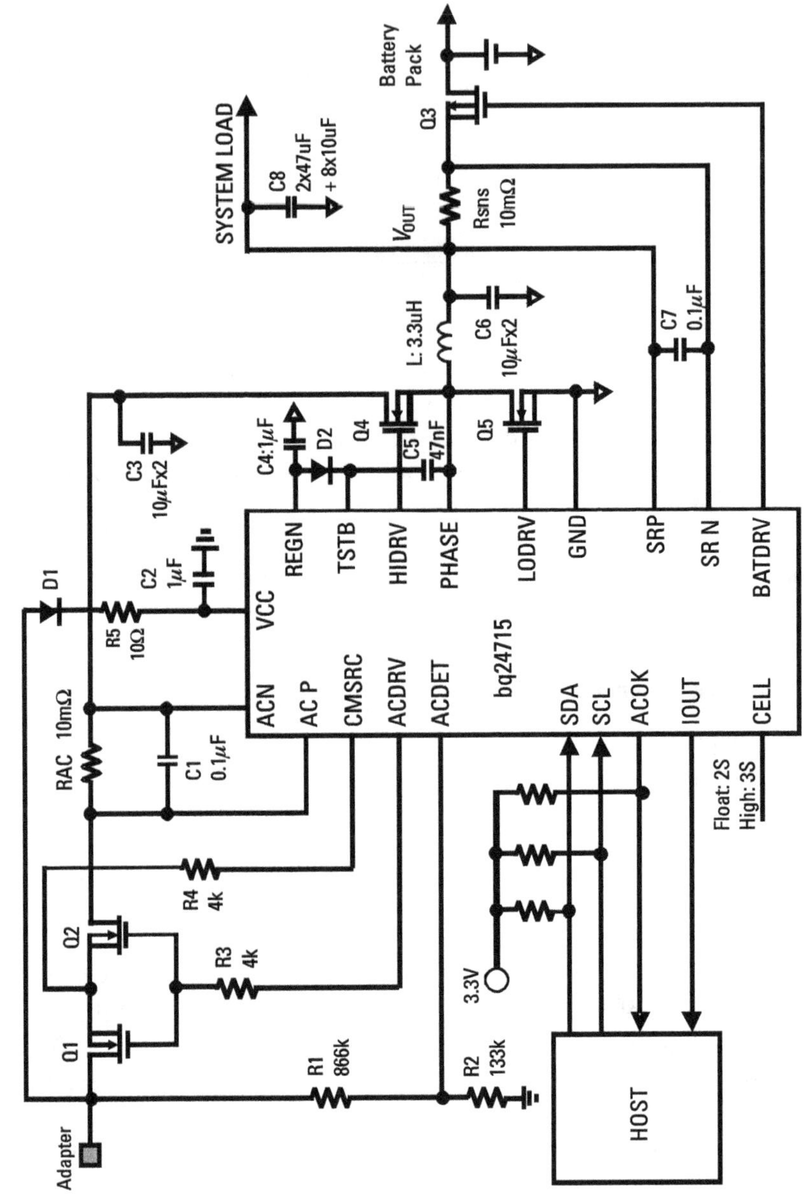

Figure 7.7   NVDC charger design example for an ultrabook.

regulating the battery precharge current, while the DC-DC buck-based charger operates in regulation to achieve minimum system bus voltage. The optimized loop compensator is integrated in the charger for achieving a fast output voltage and input current transient response during a step load system transient, for example, a CPU mode transition from the standby mode to high-performance mode. It is quite a challenge to achieve the high light load efficiency for meeting the ERP 6.0 energy star. The charger needs to be designed with a very low quiescent current of less than 1 mA and to operate in the pulse frequency modulation (PFM) mode for minimizing the switching loss.

Figure 7.8 shows the output transient response at 6V output with a deeply discharged battery. We can see that the output voltage drop is around 350 mV within 200 μs. Figure 7.9 shows the light load efficiency; it can achieve 83.5% efficiency at 180 mW of output. This provides a good design margin for achieving less than 0.5W from the adapter AC side during system standby mode.

7.6   Digital Cameras

A digital camera's battery has to provide long operation with very infrequent recharges. This requires a chemistry with a low self-discharge such as Li-ion,

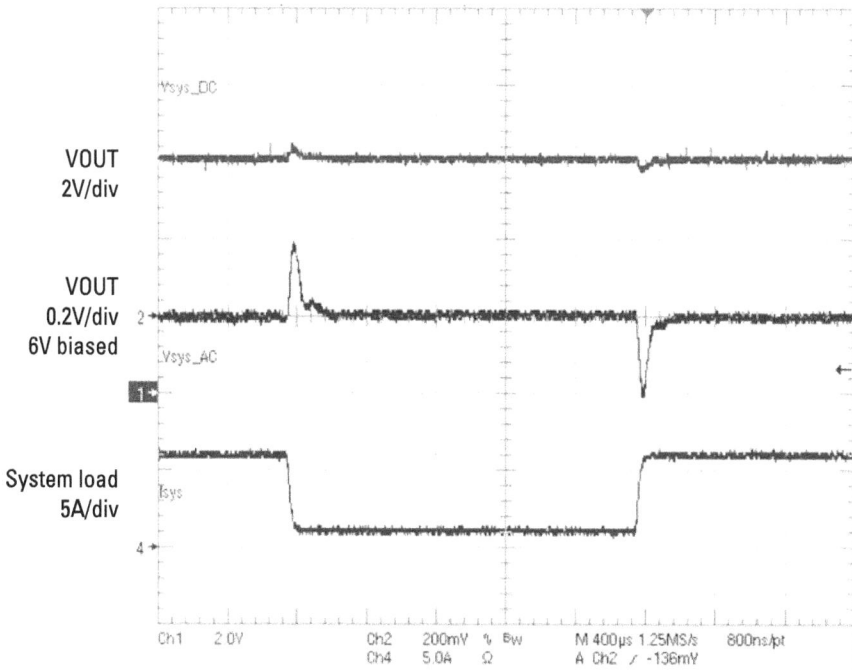

Figure 7.8   System bus step load transient response.

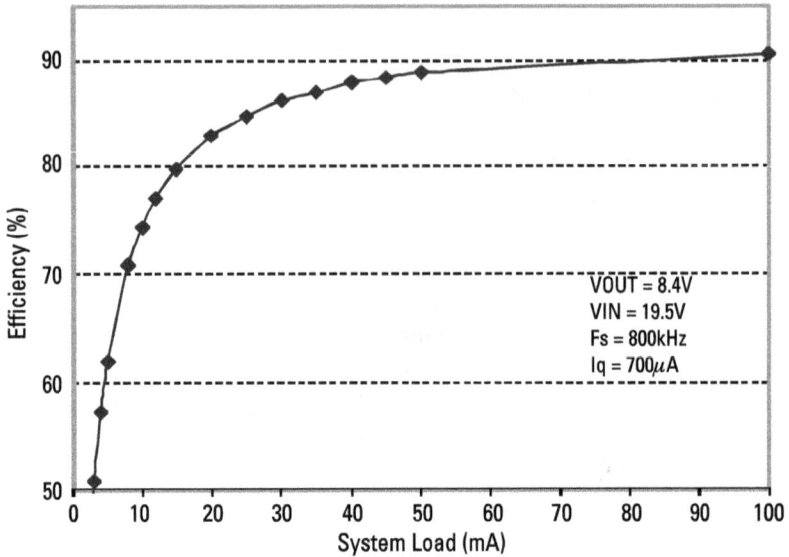

Figure 7.9   Light load efficiency.

although NiMH and primary alkaline batteries are used in some models. One important consideration is the battery pulse discharge capability, because in digital cameras most of the power consumption comes in short bursts of activity—moving the lenses to auto-focus, zoom, photo-flash (very large spike!), then readout and compress the image. A backlight is also turned on for relatively short periods of time when active picture taking is ongoing, while everything is off all of the remaining time. Figure 7.10 gives an example of power usage during digital camera operation.

It is evident that the useful operating time of a digital camera will be determined by the moment when the largest spike (usually photo-flash) drives the battery voltage below the system shutdown voltage. From this point of view, a battery needs to be tested not using just the average current of discharge, but using the actual power sequence typical for the picture taken, similar to that shown in Figure 7.10. Because battery impedance is inversely proportional to battery capacity (for the same chemistry and design), for every power sequence and system shutdown voltage there will be a minimal size for the battery that ensures any operation, just to get battery impedance low enough to take even one image without going below the shutdown voltage. For such spiky applications, it helps to have a chemistry and battery design that has lower impedance per watt-hour. For example, the recently introduced NiZn batteries have much lower relative impedance than NiMH batteries, so even with slightly lower mWh/g capacity, NiZn batteries can provide more shots per charge. Li-ion bat-

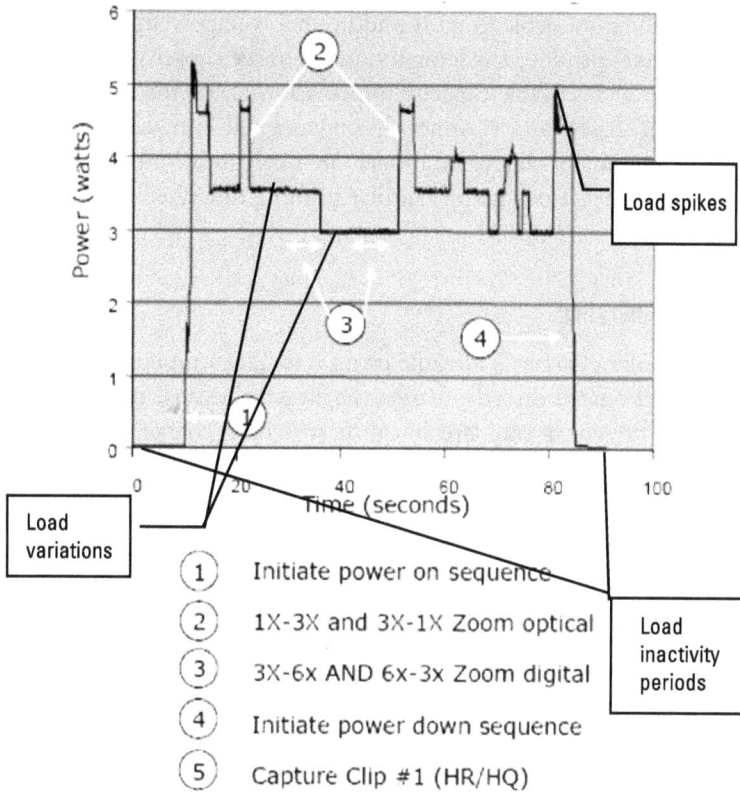

Figure 7.10 Example of power usage during digital camera operation.

teries generally have high power capability, so in most cases batteries have to be sized based on the total energy needed.

7.6.1   Battery Pack Electronics

Digital cameras primarily use a single Li-ion cell or two to four alkaline or NiMH cells. In the case of a Li-ion cell, overvoltage protection is needed as described earlier for single cell phone batteries. Battery gauging is beneficial to be able to predict useful run time. In simpler models, voltage-based gauging in the host is implemented that gives a raw, ±25% accurate indication of remaining charge. However, due to high current spikes, IR drop effects are more important in this application than in any others, so knowing just the chemical SOC derived from voltage is only of limited use. Sometimes the termination voltage is set very high (as high as 3.5V) to account for IR drop during a voltage spike. However, the IR drop is going to change with aging and at low temperatures, so it is practically impossible to set the additional voltage margin to account for all of these effects without sacrificing too much of the capacity of a new battery. A gauge that provides usable capacity information under the particular load of this application, based on continuously updated cell impedance information, would be the best and so high-end models are likely to use either host-side or a pack-side Impedance Track gauges similar to those described in the cell phone section.

7.6.2 Battery Charging

The most popular charger is a cradle charger in DSC applications, where the AC-DC adapter output directly charges the battery. Because no system interaction issues are present, a standard linear or switching charge controller can be used to ensure CC/CV charging profile.

7.7 Industrial and Medical Handheld Devices

Industrial and medical handheld devices are different from consumer devices due to their more structured usage conditions. Their daily use is often more predictable due to known shift duration, which allows users to maximize their usage and recharge the devices soon after the battery is almost fully discharged, while another fully charged battery is often available and waiting in the charge cradle. Sometimes such devices have much less downtime compared to consumer devices due to this structured, scheduled use. On the other hand, the temperature environment in which these devices are operating is often much more harsh—they might be used, for example, inside a –20°C freezer on a routine basis, so low-temperature battery testing is essential. One representative device is the portable data terminal (PDT), which is also known as a handheld data terminal. The PDT is a rugged portable computing device used in many shipping, logistics, and inventory management applications that enables greater efficiency of the mobile workforce in many industries. The main features include mobile computing with an integrated camera, a barcode scanner, and wireless connectivity to communicate and access information on remote servers. It requires an operating system that allows for custom applications, remote updates, and ease of use in the field.

7.7.1 Battery Selection

Cell selection for this application should consider low-temperature behavior. Such behavior is more dependent on the electrolyte used and cell design (such as amount of conductive additive) than on the chemistry as such. Premium cell brands such as Sanyo, Sony, LG, and SDI usually have lower impedance and better low-temperature behavior compared to less known brands. Cylindrical 18650 cells are often better for low-temperature operation compared to pouch cells because the latter often use a gelled electrolyte, which makes its conductivity worse at low temperatures. In addition, industrial devices are designed for reliability and ruggedness rather than for their looks, so there is little attempt to make them extra thin by using the more expensive pouch cells.

Longevity is another important factor for industrial applications because devices can go through a lot of charge/discharge cycles under shift operation. Sometimes use of a lesser capacity but higher cycle number of cells, as with the LiFePO$_4$ chemistry, would pay off in this environment. Overdesigning the battery somewhat would help both with longevity and low-temperature operation because low impedance will provide some space for the IR drop to increase at lower temperature and if the battery is older without causing the system voltage to drop below the shutdown level.

7.7.2 Battery Pack Electronics

The battery used in a PDT is typically a one- or two-cell Li-Ion battery depending on the voltage requirement of the particular application. If wireless capability and RF functions are needed, higher voltage packs with two cells in series are beneficial because the high current spikes associated with RF are likely to drive the voltage of a single cell below the termination voltage especially for aged cells and at low temperatures.

Safety requirements are the same as for the single-cell cell phone and notebook applications discussed earlier, and are best provided with standard gauge/protector ICs. An accurate gauging solution is essential in these types of applications because in an industrial environment overestimation of remaining charge can result in drastic productivity losses (think about a handyman who has just climbed onto a roof merely to find out that the device he is holding dies on his first attempt to drill a hole) or even life-threatening emergencies if the device is used in the medical field. Pack-side gauging is preferred because in industrial and medical settings batteries are always removable and in many cases different batteries will be used in different devices (common battery pool), so a device-side gauge will not have a chance to obtain a complete picture of a battery it has never encountered before, so it will not be able to provide accurate information. A battery pack-side gauge does not have this problem since it always resides with the battery, so information accumulated about the battery such as learned impedance and chemical capacity is preserved. Because LED functionality is very useful in this application, multiple-cell packs often use notebook-type gauges with LED support such as that shown earlier in Figure 7.4.

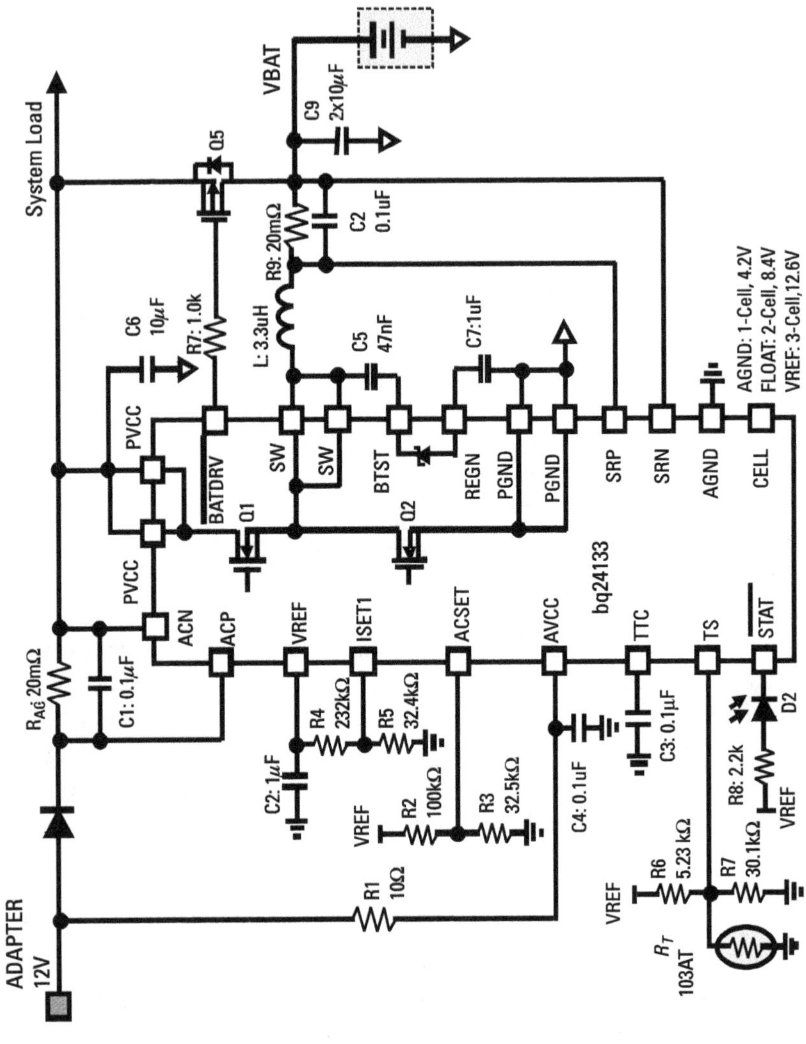

Figure 7.11 Two-cell Li-ion battery charger for PDT applications.

7.7.3   Battery Charging

In a low-end application, a low-cost linear battery charger is usually used. But in medium-and high-end applications, a high-efficiency switch-mode charger is used. For an application with a printer driven by a motor, a two-cell Li-ion battery pack will be used because a certain amount of voltage and peak power are needed in order to efficiently drive the motor. Figure 7.11 shows the switch-mode charger with an integrated power MOSFET for charging a one- to three-cell Li-ion battery application with a 2A fast-charge current, 200-mA precharge current, input current regulation of 2A, and a 9-hour safety timer.

7.8   Conclusion

Regardless of how you arrived at this place, either by reading the entire book or by jumping from place to place to locate needed information, we hope you were able to find what you were looking for. As promised, we have traveled through the sparkling realm of battery chemistry, stopped in the fiery cave of battery safety, got recharged in the charging section, wandered through the woods of cell balancing, touched on the mysteries of battery gauging, learned some surprising facts about ESD and EMI, and finally picked through a collection of portable device examples that we hope allowed you to examine how all of the accumulated experiences can materialize into a delightful electronic toy for kids and adults alike. We thank you for making the journey with us, and may the battery power be with you always!

About the Authors

Yevgen Barsukov is an IP development manager in the battery management systems group and a distinguished member of the technical staff at Texas Instruments, Inc. (TI). Dr. Barsukov specializes in applying leading theoretical methods of battery analysis to improve the battery control technologies used for fuel gauging and the safe and healthy operation of notebooks, mobile phones, PDAs, and other portable devices. During more than ten years at TI, his work on the development of widely adopted industry gauging methods, such as Impedance Track™, contributed to TI earning a leading market share position in the area of portable power management. Prior to joining TI, Dr. Barsukov's research was focused on impedance spectroscopy testing and modeling of batteries. He has presented papers at numerous international conferences and workshops, written for journal publication, and coauthored the book *Impedance Spectroscopy*, Second Edition, which is considered a standard reference in the field. Dr. Barsukov earned an Ms.C. in organic chemistry at Kiev National University in 1993 and a Ph.D. in physical chemistry from Kiel Christian-Albrecht University in 1997.

Jinrong Qian is a product line manager for battery charge management and an emeritus distinguished member of the technical staff at Texas Instruments. He is a pioneer in developing innovative AC-DC and DC-DC power converters and state-of-the-art battery management technologies for significantly improving battery charging efficiency and safety and extending battery run times. He holds 26 U.S. patents in power management and has published more than 75 professional technical articles in the area of power management. He served as an associate editor of *IEEE Transactions on Power Electronics*, and is often invited to give professional tutorials and seminars in power management at the IEEE Applied Power Electronics Conference and Battery Power Conference.

Mr. Qian was named the 2011 Asian American Engineer of the Year by the Chinese Institute of Engineers for his exemplary contributions in science and engineering to the semiconductor industry and in recognition of the positive impact he has made on the Asian American community. Mr. Qian earned his Ph.D. in electrical engineering from the Center for Power Electronics Systems at Virginia Polytechnic Institute and State University in 1997, an M.S. in electrical engineering from the University of Central Florida in 1994, and a B.S. in electrical engineering from Zhejiang University, China, in 1985.

Index

www.ingramcontent.com/pod-product-compliance
Lightning Source LLC
Chambersburg PA
CBHW050456190326
41458CB00005B/1311